Social and Ecological Systems

Association of Social Anthropologists

A Series of Monographs

A.S.A. MONOGRAPH 18

Social and Ecological Systems

Edited by

P. C. BURNHAM

Department of Anthropology
University College London
England

and

R. F. ELLEN

Eliot College
University of Kent at Canterbury
England

1979

ACADEMIC PRESS

London   New York . San Francisco

A Subsidiary of Harcourt Brace Jovanovich, Publishers

ACADEMIC PRESS INC. (LONDON) LTD.
24/28 Oval Road,
London NW1

United States Edition published by
ACADEMIC PRESS INC.
111 Fifth Avenue
New York, New York 10003

British Library Cataloguing in Publication Data
Social and ecological systems.
 1. Man – Influence of environment – Case studies
 I. Burnham, P II. Ellen, Roy III. Association of
 Social Anthropologists of the Commonwealth.
 Annual Conference, Cambridge, 1978 IV. Social
 Science Research Council (Great Britain)
 301.31   GF51   79–50520
 ISBN 0–12–146050–9

Printed in Great Britain by
John Wright and Sons Ltd., The Stonebridge Press, Bristol.

PREFACE

The 1978 Conference of the Association of Social Anthropologists departed from tradition in being a joint venture organised with the British Social Science Research Council. We would like to express our gratitude to the Council for its valuable support, and also to the British Academy, the British Council and the Maison des Sciences de l'Homme for enabling us to invite overseas guests. However, it should be made clear that responsibility for the content, opinions and conclusions expressed in this monograph rests entirely with the authors and not with these funding bodies.

At the invitation of Barbara Ward, the venue of the meeting was Newnham College, Cambridge, whose hospitality was both flawless and generous. This was very much due to the efficient liaison arrangements made by Carol MacCormack. We would like to thank those who, in addition to ourselves, convened the various sessions — James Woodburn, Emmanuel Marx, John Harriss and Carol MacCormack, and those who presented papers which are not represented in the present volume. We are also grateful to Richard Fardon for his help with the index.

CONTENTS

In memory of Daryll Forde

INTRODUCTION:
ANTHROPOLOGY, THE ENVIRONMENT AND ECOLOGICAL SYSTEMS[1]

ROY F. ELLEN

There is no consensus on what constitutes an ecological approach. At times the quite erroneous impression is given that we are somehow dealing with a coherent methodology which has developed in response to a clearly-defined set of empirical and theoretical problems. This is clearly not so. Over a period of one hundred years the man-environment problematic has been successively re-formulated, and old approaches have given way to new ones. But many old ideas have never been entirely rejected, and new ideas are seldom completely accepted. Moreover, many anthropologists are apparently unaware of the complex intellectual history forming the backdrop to the issues they investigate. As a result, a contemporary 'ecological approach' may refer to a particular version of systems-analysis (particularly the study of self-persisting systems), an interactionist model derived from the concept of ecosystem, causal hypotheses of a 'cultural materialist' type, or simply any study which addresses itself to the interrelationships between social and environmental variables with a modicum of sophistication. In addition, there are now textbooks, modules, readers, review articles, programmatic statements and entire journals devoted to it.[2] The papers collected here reflect such a diversity of views, though not always approvingly; but the area of common ground is still considerable.

It would be an unnecessarily repetitious and space-consuming exercise to summarise all this here. However, given the auspices under which this monograph appears and the academic background of most of the contributors, it is highly appropriate that we should at least consider briefly the position of ecological relations in the British anthropological tradition, which has been curiously peripheral to mainstream trends. Beyond this I make some preliminary observations on the place of man-environment studies and the various contemporary ecological procedures and approaches, as they relate to the contributions in this volume.

Environmental Relations in the British Tradition

Because the examination of the connexions between environment and social behaviour has always been closely associated with geographical determinism, what is now described as *ecology* has had an ambiguous position in British anthropology. In so much as the work of our Victorian predecessors was evolutionist, racialist, sociological, historical, influenced by Hume and the romanticists or the Benthamite school, so it was implaccably opposed to geographical determinism. But as a product of the Enlightenment it was also residually environmentalist. Comte, who was so influential in shaping a later social anthropology, though critical of the environmentalism of Montesquieu, was himself prone to exaggerate the efficiency of physical agencies. Although Ferguson strenuously upheld that the passionate ingenuity of man enabled him to overcome the disadvantages of any climate or situation, he was still of the opinion that temperate zones facilitated the growth of civilisation. For Herbert Spencer, history was the steady progression from warmer and more productive habitats demanded by the feebler stages of social evolution to the colder, less productive and more difficult regions away from the tropics. Morgan held that the cultural differences of peoples at the same 'level' were to be explained environmentally. Over and above this there was always (a sometimes implicit) folk-environmentalism affecting the often slipshod and untutored judgements of travellers, missionaries, colonial officers and other early ethnographers. So perhaps environmentalism was not as incompatible with then current racist and evolutionary theories as is popularly thought.[3]

But at the level of theoretical and ideological generalisation, the cards were stacked against environmentalist interpretations. The theories of Le May, Ratzel and Demolins were, by general consent, 'Far too pretty to be true' (Marett 1912: 97−100). Such suspicion was reinforced by theoretical developments in the first half of the twentieth century. Neither diffusionists nor functionalists could tolerate the environment as an element in causal processes if their theories were to remain elegant and free of embarrassing contradictions. Comparable and even more specific objections to environmentalism were meanwhile being articulated in the United States by Franz Boas and his students, for whom history seemed to provide much more satisfactory explanations.

Moreover, the extent to which a society was able to modify its surroundings was becoming increasingly clear: no longer a matter of philosophical assertion but of ethnographic confirmation. In such circumstances empiricism was a safe doctrine. Detailed evidence was thus marshalled against the various versions of environmentalism and materialism, including a caricature of Marxism. It hardly helped matters that determination by the environment was sometimes confused with that by the technical means of subsistence and production.

That an emerging British school of social anthropology should have adopted so apparently intransigent a position is no doubt partly due to the necessity of establishing a set of distinctive intellectual credentials, but it was also a result of internalising the circular explanations fashionable in French sociology. It is noteworthy that Durkheim's principle of only explaining the social in terms of the social closely paralleled the Boasian notion of 'superorganic' in its epistemological and polemical function; society like culture was a thing *sui generis*. Furthermore, the Durkheimian slogan underpinned the subsequent claim that the discipline be treated as an autonomous source of explanations (Gluckman 1964).

Theoretical developments, then, discouraged the treatment of environmental relations, and they rarely featured at the level of generalisation, occupying a peripheral and passive role.[4] Radcliffe-Brown, in the words of Netting,

> consigned studies of subsistence adaptation to an 'external realm' where labored archaeologists and ethnological historians of diffusion. The topic could safely be left with museum catalogers of material culture or such contemporary experts as the geographer, the agronomist, and the rural sociologist. (1974:21)

The same attitude is reflected in later textbooks. Theoretically, ecology remained residual; the dominant paradigm was one that ruled out a central place for man-environment interactions.

Among the first two generations of British social anthropologists there was one man who maintained virtually single-handed a theoretical interest in environmental relations, although at the same time he was also responsible for articulating the case against geographical determinism. This was C.D. Forde, whose position is also commented upon by Paula Brown in her paper.

Forde had been trained first as a geographer and then as an archaeologist before turning to social anthropology. As a geographer he was aware of the crude environmentalist assumptions still current in that discipline but had been particularly influenced by the works of L. Febvre, P. Vidal de la Blache and R. Ahrens. As an archaeologist and anthropologist who had received part of his training in America, he had been affected in part by the historical particularist approach of Boas, Kroeber and Lowie, but equally by the empiricist British ethnographic tradition. He rejected the culture area concept on the grounds that general classifications are inadequate for the analysis of cultural possibilities and because local histories often showed it to be of little applicability (Vayda and Rappaport 1968: 482). Unlike the Boasians though, who were largely concerned with linguistic texts, Forde concentrated more on material culture and environmental data. His first and most important book, *Habitat, Economy and Society*, first published in 1934, was a general treatise with comparative ethnographic sketches which are ecologically-oriented. In it he is consistently concerned with functional relations, both internal and external to the particular society being discussed. He shows continual awareness of alterations in the value of environmental variables due to technological innovation and social change and the complex inter-weaving of synchronic connexions and historical developments (Alfred Harris in Fortes 1976: 473). Perhaps the key passage in understanding his general position is the following:

> Physical conditions enter intimately into every cultural development and pattern, not excluding the most abstract and non-material; they enter not as determinants, however, but as one category of the raw material of cultural elaboration. The study of the relations between cultural patterns and physical conditions is of the greatest importance for an understanding of human society, but it cannot be undertaken in terms of simple geographical controls alleged to be identifiable on sight. It must proceed inductively from the minute analysis of each society. The culture must in the first place be studied as an entity and as an historical development; there can be no other effective approach to interrelations of such complexity. (1934: 464—5)[5]

The position of Forde on environmental issues was essentially a product of his eclectic theoretical socialisation, but for those more firmly entrenched in social anthropology, the experience of field-work was to play a key role in determining attitudes.

The environment could not be ignored, for it was the first thing which confronted the fieldworker. It was widely regarded that 'no sociological study of a community could be undertaken without an understanding of the natural environment within which it exists and from which it draws its subsistence' (Notes and queries 1951: 35). And yet it was obvious that environment rarely simply determined social organisation. It was therefore easy to resort to generalisations claiming that the environment acted only to limit what was possible.

Given a certain *a priori* theoretical disposition, the treatment of ecology (generally meaning 'environment') was seen as obligatory, but in the event something of a ritual exercise. It became a discrete section in monographical studies, often the first chapter; a few background facts on the distribution of vegetation, rainfall records and topographical features were presented *in vacuo* before proceeding to the main focus, the analytically autonomous domain of social organisation. The most important and virtually only exception to this picture until the mid-fifties was the work of Audrey Richards on the Zambian Bemba (1939).

Nevertheless, fieldwork experience and the analysis of ethnographic data compelled a softening of polemical attitudes. As is often the case, individual analyses tended to push the ethnographer into a more moderate position, towards a stress on the substantive uniqueness of his or her own data. If this was the general case, then on occasions the specific conditions of fieldwork demanded a particularly close attention to environmental factors. For example, ecological models appear much more successful in explaining hunter-gatherer behaviour. Such societies are often small, dispersed, self-sufficient, modify their environment to a limited extent only, but perhaps most importantly, exhibit social differentiation such that group behaviour tends to express the cumulative and consensual adaptations of individuals. Revealingly, when ecology does at last get a mention in a widely-used textbook (Mair 1972), it is in relation to a discussion of hunting and gathering. In different ways, fieldwork in societies with other subsistence bases has had similar results; among pastoralists, for example, and expansionist swidden cultivators. In such cases environmental relations are superficially represented as being clearer — more direct.

Partly as a result of the influence already mentioned, but also in part as a response to the explanatory sterility of pre-war functionalism, the fifties and sixties gave rise to a series of studies which while broadly phrased in the language of 'possibilism' made some attempt to explain patterns of structural variation in terms of key environmental variables. Two works which seem to have been instrumental in stimulating this trend are Leach's *Political Systems of Highland Burma* (1954) and Worsley's critique of the Fortes analysis of Tallensi kinship (1956).

Among the more important of these studies were a series undertaken on East African pastoralist societies. It is interesting and significant that Forde singled out a number of these for discussion in his 1970 Huxley memorial lecture. He examined two examples, based in the first place on Evans-Pritchard's Nuer data (1940), and in the second place on the analyses of Gulliver (1955) and Dyson-Hudson (1966) on the Jie, Turkana and Karimojong of northern Uganda and Kenya. It is useful to briefly summarise the main arguments here to illustrate the dominant character of ethnographic analyses concerned with environmental relations in the British tradition. Their plausibility is consequently only of secondary interest.

In his analysis of Nuer political relations, Forde argues that relations between larger Nuer tribes not only reflect the process of territorial expansion and demographic growth but also the underlying differences in environmental conditions. He is able to do this by observing the different form of political relations in two separate environmental zones. In the earlier Nuer homeland west of the Upper Nile, the topography offered only small and isolated areas of high ground for seasonal occupation. Seasonal movements tend, therefore, to be confined. Tribal units are correspondingly small, around 10,000 persons. On the other hand eastern Nuer land is characterised by more extensive areas of higher ground. The wet season settlement area is correspondingly larger and water supplies are more restricted. This leads to a much wider dispersal of the population and larger political groups of between 50,000 and 60,000 persons (cf. Bonte, this volume).

The second example is concerned with the interpretation of the different kinds of local and formal groups found in the Karimojong cluster, composed of the Jie, Karimojong and Turkana. The Jie and Karimojong camp around river-bed waterholes during the dry season: all that is available. Population density is around 15 persons per square mile. Homesteads are built permanently at reliable water points by agnatically related kin, usually groups of brothers. Some cultivation is practised. Among the Karimojong, agriculture is rather more marginal and there is less stress on agnatic ties. The country occupied by the Turkana is much more arid with a population density of around 3 to 4 persons per square mile. The uncertain climatic conditions are allied to the absence of concentrated or permanent settlements. Cultivation is absolutely minimal depending on irregular rainfall. The herd owning unit is correspondingly smaller and is never more than two generations deep. Here there is a greater reliance on an extensive and dispersed set of social relationships. So, as we move from the Jie to the Karimojong to the Turkana, there is increasingly earlier fission of agnatically related groups correlated with an increasingly uncertain and arid environment.

These summary discussions of the literature may be misleading in that they are, necessarily, contractions of extended and often intricate analyses. But they do neatly show the pervading character of such studies — broadly possibilist and correlative. In many ways they are the sociological analogue of twin studies, 'the exposure of the same or similar cultural materials to different environmental influences' (Vayda and Rappaport 1968: 487).

There is no denying that the British possibilist tradition of analysis of the relationships between environment and social structure, despite being wedded to sociological functionalism and ethnographic empiricism, has genuinely contributed to the development of human ecological studies in several important ways. It has fostered exploration of the concept of 'limiting factor'. It has been linked to an acknowledgement that physical conditions intimately enter into social and cultural patterns — that there is an intricate relationship between environmental and social variables. It has given increasing recognition to the variation and extent to which man has modified his environment. It has warned against simple geographical controls and advocated the detailed analysis of each society. The focus on detailed empirical fieldwork, however, has been the most important legacy of the possibilist viewpoint, and it has been fieldwork which encouraged and sustained its general theoretical stance.

Where possibilism succeeded was in its stress on the importance of empirical investigation; where it was inadequate was in its methodology and theory. Because it was linked in part to anti-Marxist and anti-materialist ideology and the rejection of evolutionary holism, it was compelled to adopt a sometimes crude inductionism. Its empirical strength was reduced to empiricist weakness. Anthropologists were increasingly bewildered by the variety of cultural experience. Possibilism addressed itself, therefore, to limited problems cast in an ostensibly scientific idiom but detached from questions of causality and origin and other parts of ecological systems in which they were enmeshed. The problem was that by using the safe phraseology of vaguely permissive and restrictive parameters of 'constraints' and 'limiting factors' it believed it had solved the problem of determinism. It had not. The environment does indeed set limits on what is socially possible under a given set of conditions, but in restricting the options it is obviously also helping to determine the final outcome.

Ecological Functionalism and Systems Ecology

The fifties and sixties, then, gave rise to a more sensitive appreciation of the relations between social behaviour and environment in the ethnography of social anthropologists working within the British tradition, although it was still broadly possibilist and empiricist in character. Gradually, though, certain changes were taking place. Particular ethnographic analyses showed a new subtlety in the handling of man-environment interaction (Barth 1956) or crossed disciplinary frontiers (Brookfield and Brown 1963). Most importantly, as part of a growing discontent with structural-functionalism as a theoretical orientation, the idea that social structure or culture was a closed system came in for severe criticism (Gray and Gulliver 1964: 11; Gellner 1964). Although initially the developments in American anthropology, the cultural ecology of Steward, the materialism of Marvin Harris, went largely unnoticed in Britain, the cumulative weight of the literature in these fields eventually had its effect, at least on the younger generation of social anthropologists emerging in the late sixties and seventies.

Under the impact of systems theory and biological ecology, functionalism itself took on a new lease of life. Behavioural and environmental traits were viewed as parts of a single system; the two-way nature of causality was acknowledged. The new analytical framework permitted the observation of new kinds of relationships, led to the formulation of fresh hypotheses, posed old problems in new ways, and drew attention to the wide range of ecological variation. Ecosystems (which could be specified as either generalised or specialised) became the new analytical and comparative units. Moreover, they were seen as examples of general systems and therefore seen as possessing rational structure and regulatory properties; they were conceived of as flows of energy, materials and information (Vayda and Rappaport 1968; Ellen 1978).

So, at least in terms of British social anthropology, man-environment relations have at last been brought fully and acceptably within the ambit of social theory. But some writers have shown so much enthusiasm for the new ecology that occasionally the impression is given that we have now reached an epistemological nirvana — that, if only the rules are applied correctly, an ecosystem approach will explain everything (see, for example, Anderson 1973). Fortunately, the new faith already has its heretics (e.g. Friedman 1974), and the papers collected here generally reflect a revised or openly sceptical stance concerning some recent trends. This is not to say that they are therefore necessarily negative; indeed, they build on concepts, theories, methods and data drawn from studies which in other respects they might be expected to be critical of. There are repeated reminders that we should never underestimate the ingenious and pre-eminent role of social (including historical) factors as causal agents, and that social and environmental variables interact in subtle and complex ways, which are just not apparent from simple correlative studies. Accordingly, the dissection of relationships between them requires enormous care and appropriate skills.

It is impossible in an introduction of this kind to do justice to the full range of issues tackled in the various contributions to this volume. Instead, I focus on just two related themes which the contributors to this monograph refer to: matters which have already occasioned discussion in the literature. These themes are: (1) the problem of analytical closure or boundedness in the study of social systems seen in their wider regional context, and (2) the difficulties inherent in the elegant but conceptually inadequate definitions of social adaptation, particularly in connection with cybernetic models of social process. Both require careful examination if we are to escape from naive ecosystems ethnography.

Variation, Comparison and the Boundaries of Explanatory Units

Analytical closure poses a particularly acute problem for systems analysis, although it is by no means unique to it. The search for 'systemness' rather than arbitrary definitions remains an important issue in the interpretation of human behaviour patterns in general (Buckley 1967: 42). It is practically and intuitively relatively easy to isolate 'systems' which make sense of data and which may be used for the investigation of a wide range of different research problems. However, it is empirically complicated to specify them in terms of numbers of connections or intensity of flow. The boundaries of systems may often be modified from an original *a priori* delineation, so as to conform with emerging data, or to make data conform with a model. Moreover, as Jonathan Friedman points out, the difficulty of guaranteeing closure under empirical conditions is that local societies are rarely closed reproductive units (see also Friedman 1976). Where closure is demonstrably weak, local societies are no longer sufficient units of explanation.

These issues are illustrated ethnographically in the papers by Paula Brown, Roy Ellen, James Fox and David Harris.

Harris is dealing with three local but inter-insular systems, which despite important external trading connections, exhibit a considerable degree of functional and reproductive autonomy. Each small 'community' has a similar ecologically-related division of labour and pattern of exchange, whereby small islands with intensive horticulture are the foci of larger 'peripheral' areas characterised by food-collecting.

A similar pattern seems to have been prevalent in the Moluccas for the historical period examined by Ellen. But here trading links of a regional and increasingly global kind seem to be emerging as determinate in explaining the reproduction of local political and economic structures. Small local kinship units, theoretically specified as self-reproducing, are progressively drawn into increasingly wider exchange networks, and larger and more complex modes of production and systems of total reproduction. The process by which this happens is closely related to the properties of *Metroxylon* sago as a subsistence resource which facilitates expansion in the spice trade over the course of several hundred years.

Not only do regional studies pose important questions about the level at which systems are able to reproduce themselves but they also illuminate social and ecological variation which allows us to evaluate particular hypotheses concerning local populations. Paula Brown illustrates this for the New Guinea highlands, as does David Riches for the American Northwest coast.

The existence of variation invites the investigation of controlled comparison, a theme developed by many of the contributors. Alan Barnard's paper is a careful and detailed inventory of significant variation for Bushman groups in the Kalahari region, which casts some doubt on the typicality of some individual analyses (such as those of the !Kung) and wider generalisations based on them. Fox examines the ecology surrounding the use of lontar palm sugar as the basis of economic and social variation in the historical development of two Rotinese states. Bonte is particularly concerned with variation in segmentary lineage organisation under differing ecological conditions and, by comparing a number of pastoralist and agro-pastoralist societies, is able to show that although divergent ecological relations go some way in explaining differences, they are insufficient in themselves. Moreover, by viewing sets of Nilotic and Bedouin societies in both comparative and historical perspective, he develops arguments against the classic equilibrium model of segmentary lineage societies and points to the importance of economic and political inequalities in the emergence of pastoralist hegemonies.

Adaptation, Change and 'Environmental Fit'

The problem of closure, and the related issues of variation and comparison, lead on easily to the second major theme, adaptation. This is because any claim that social behaviour is adaptive is always accompanied by assumptions about the level at which it is so: the individual, kin group, social class, local community, polity, society, or — for that matter — regional system. While certain practices may well be 'adaptive' in the sense that they contribute to the reproduction of the system, the system that is being maintained may be to the immediate disadvantage of certain groups and individuals within it.

It may also be to the long term disadvantage of those who benefit from the reproduction of the social structure as a whole because of internal contradictions which only work themselves out over time. Hence Friedman's question: 'Adaptive for whom?' Closure is also a precondition for establishing the presence of negative feedback, and therefore, because it is difficult to guarantee, it becomes a key obstacle in all such models of adaptation. Three of the papers — those by Brown, Burnham and Friedman — pursue critically the notion of negative feedback as developed by Roy Rappaport.

Brown is specifically concerned with the applicability of Rappaport's model to his own Maring material and to other New Guinea highland societies with similar means of production, exchange and social structure. She is critical of assumptions that European contact disturbed what had hitherto been equilibrium systems and illustrates this with reference to her own studies with Harold Brookfield of pig-raising, feast frequency, colonial and pre-colonial change among the Chimbu. In this she is contributing to a continuing debate in the Melanesian literature and elsewhere (e.g. McArthur 1977; Vayda and McCay 1977). Following Friedman, she emphasises diversity, competition and expansiveness as parts of a situation of dynamic non-equilibrium.

By emphasising the asymmetry introduced into areas through historical change, the papers by Fox and Ellen also cast extreme doubt on the existence of simple equilibrium systems in the pre-colonial and early colonial periods of eastern Indonesia. Indeed, one of them goes some way towards documenting those non-homeostatic processes of positive feedback, of which our knowledge is still so lamentably incomplete.

Whereas Brown's contribution is mainly an ethnographic critique of Rappaport, Friedman's is very definitely in the realm of abstract theory. Not only does he repeat some of his well-known views on ecological functionalism and cultural ecology (Friedman 1974) but he casts his net wider to include his sometime mentor Maurice Godelier and Gregory Bateson. Friedman explicitly links Rappaport's approach to the assumptions of Bateson's Hegelian concept of 'Mind'. Both, he insists, are concerned with the idea that culture is a well-ordered cybernetic totality, where everything which is destructuve or contradictory is incorporated into a model of harmony. Such ideas for Friedman are suspiciously convenient ideological props for utopian environmentalists in the modern industrial age (cf. Cotgrove 1976). This line of argument is similar to Burnham's point that Rappaport all too easily falls back on that familiar myth, 'the ecological wisdom of the actors'. For both Burnham and Friedman such false assumptions find their apogee in notions of control hierarchy as applied to social systems, where religious beliefs represent the highest order regulator and where defining religion as adaptive reveals 'a clear and somewhat disconcerting ideological bias' (Friedman, this volume).

For Tim Ingold, adaptation is a temporary condition. In the most general terms, this is also Friedman's position, but he claims that in practice social systems have never been adaptive and that their internal dynamics have always tended to be of an accumulative nature. Burnham too resists the use of the term 'adaptation'. But he believes that there is something to be salvaged from the notion of negative feedback if it is used simply as a description of a structural state of affairs and is shorn of the implication of purposive cybernetic regulation. As an ethnographic case in point, Burnham contends that for the Gbaya of east central Cameroon, cease of residential mobility is a major element in 'triggering' a causal cycle which results in the maintenance of a complex conservative 'structural nexus'. It is this process that for Burnham relieves pressure on individual autonomy which might otherwise lead to structural transformation. He sees residential mobility serving as a regulator but, rather than being the end product of a process of adaptation, it is no more than the effect of a par- ticular arrangement of variables within given limits. Burnham, like Friedman, draws our attention to the recent emergence of the field of 'stability theory' (Holling 1973; Glansdorff and Prigogine 1971), which offers an alternative to cybernetic explanations of structural stability.

Peterson's paper, on the other hand, comes down firmly on the opposite side of this debate, arguing that territorial organisation among desert Australian hunter-gatherers is a cultural encoding of resource distribution which ensured an effective adaptation of population size to resource base in these groups. Pursuing these ideas further afield, Peterson argues that similar cultural mechanisms are also to be found among Bushman groups and concludes that cultural materialists' overemphasis on observed behaviour has tended to obscure the important adaptive role of ideology.

Van Leynseele and Riches are also interested in the fit between environments and social practices, without attributing the property of adaptation to the systems which they discuss. Van Leynseele's paper is an exercise in what is sometimes called 'ethnoecology'. He shows the way in which technical practices are essential for the stability of a specific ecosystem, the vast floodlands of the Congo-

Ubangi confluence in Zaire. He describes how the knowledge necessary to sustain the system is reflected in Libinza classifications of environmental space, ecological successions and seasonal processes, including the movements of fish. Overall, a number of Libinza social and technical practices have the effect of preventing overfishing, but when this does occur, Van Leynseele indicates, there are well-understood ways of curbing appropriation and redistributing population.

Riches argues that environmental constraints may be crucial to making variation in the pre-colonial social structures of the American Northwest intelligible; for as we move northwards so flexible notions of ownership, rank and kinship acquire a much greater rigidity. Riches would maintain that different ideologies embody strategies which actors may employ to control manpower under varying environmental conditions. So, as with Van Leynseele, actors are not seen as passively reacting to environmental change, or as locked into mechanistic systems of negative feedback, but actively and consciously manipulating their social relations to their perceived advantage.

Ingold has an altogether more positive view of adaptation than most other contributors, one that is strictly Darwinian. However, in common with his fellow contributors, it seems that he would reject the application of the concept to *systems*, reserving it for *practices*, such as patterns of cooperation, skills, organisational techniques and knowledge. Whether or not these function as adaptations is open to dispute, but this is very often their effect. So, while group selection and adaptation may be theoretically difficult to account for and empirically awkward to describe, there is not the same problem in suggesting that individuals adapt to changing environmental circumstances, though they may do so through the manipulation of social relations, through cooperation, exploitation or conflict (Ruyle 1973). It seems that we can only speak of the adaptation of groups in a derivative sense, that it is the product of either cumulative adaptation of individuals or the manipulative adaptation of powerful individuals or collectivities within a group (Ellen 1978a).

Ingold's approach begins from a point which for him is epistemologically axiomatic — that as part of nature, human culture must be subject to laws governing all living things. To this extent, he is sympathetic with Rappaport's position and is a modern representative of a long tradition of anthropological theory to which many would subscribe. But he argues that the conjunction of social and ecological relations at a particular point in time defines the set of conditions to which organisms are constrained to adapt, culturally, genetically, or through a combination of the two. For Ingold, cultural adaptation is a much more sensitive means of change, because it is a relatively immediate and conscious response. Thus, it permits a population to survive an ecological transformation and enter a new phase of social, as distinct from ecological, evolution. Genetic adaptation permits the intensification of economic production beyond the limits of the functioning of the ecosystem and therefore necessarily causes a demise of the population, thereby permitting a temporary restoration of equilibrium, defined strictly in ecological terms. His argument is a subtle one and should not be read as simply further naive biologising of man. In fact, his intention is expressly the opposite; he exhorts us to study the socialisation of animals.

Social and Ecological Systems

Such criticisms touched on here by no means apply equally to all analyses employing a systems approach, but it is now evident that the faith placed in it has sometimes been misguided. While the notion of system is a fundamental one and has led to significant conceptual and empirical advances, it is insufficient in itself. The selection of variables, the degree to which they must be divided, conveniently grouped or ignored, and the level of closure for any analysis, are all matters which must be determined by a particular theoretical orientation. The notion of a 'system' is incapable of assigning relative importance to variables and relationships, and when this is done it must be based on assumptions of causality extrinsic to the idea of system (MacCormack 1978).

Social organisation is not some kind of reified whole through which all environmental interactions are channeled. The environment affects individuals and individual social relations. Social systems are to a large degree open and interact through numerous links with environmental variables. But this does not mean that they do not also have an internal systemness which must be understood on its own terms. They are not simply reducible to energy flows; production is not restricted to calories, or even other materials, but concerns value as well (Cook 1973; Ingold, this volume).

At its most general level, this collection of twelve papers throws into relief the articulation of social and ecological relations under a number of specific ethnographic conditions and also at the level of theory. In doing so, it emphasises that the ecology of human systems is not to be understood as an unproblematic extension of that of lower organisms, and that ecology cannot itself provide us with a theoretical framework without also anticipating established social and economic theory. The 'ecological approach' does not swallow-up its more conventional alternatives but complements them.

On the other hand, because earlier versions of the man-environment problematic have been discredited through the discovery of some critical weakness, this does not mean that ideas and empirical research inspired by them have nothing left of value. Ecological and ecosystems approaches will continue to be attractive because in many cases they are a useful — if not essential — aid to analysis. Despite the fact that systems seldom allow for adequate problem formulation, let alone problem solving, in themselves, ecology provides a range of technical procedures, data, explanatory and organising concepts which cannot be ignored.

Footnotes

1. It would be quite improper for me to proceed without acknowledging the close cooperation and good advice of Philip Burnham in writing this introduction.
2. E.g., Anderson (1973); Bennett (1976); Ellen (1978); Hardesty (1977); Netting (1971); Vayda (1969); Vayda and Rappaport (1968).
3. See Thomas (1925); Lowie (1937: 58) and Stocking (1968: 251).
4. As James Fox aptly reminds us in his paper, environmental relations were not entirely neglected by the Durkheimians, and Mauss and Beuchat's monograph is quite rightly held to be a classic of its kind. But then it becomes all the more curious as to why it has been neglected.
5. Despite his insistence that "broad general classifications of climatic and vegetational regions are quite inadequate for the analysis of cultural possibilities" (1934: 464), Forde certainly over-emphasised the homogeneity of the three vegetation zones of West Africa in his analysis of the development of states in that region (1953; cf. Morton-Williams 1969: 80).

References

Anderson, J.N. 1973. Ecological anthropology and anthropological ecology. In *Handbook of social and cultural anthropology*, J.J. Honingmann (ed.). Chicago: Rand McNally.

Barth, F. 1956. Ecologic relationships of ethnic groups in Swat, North Pakistan. *Amer. Anthrop.* **38**, 1079—89.

Bayliss-Smith, T. 1978. Maximum populations and standard populations: the carrying capacity question. In *Social organisation and settlement: contributions from anthropology, archaeology and geography, Part I*, D.R. Green, C.C. Haselgrove and M.J.T. Spriggs (eds.). (B.A.R. International Series [supplementary] **47i**). Oxford: British Archaeological Reports.

Bennett, J.W. 1976. *The ecological transition: cultural anthropology and human adaptation*. Oxford: Pergamon Press.

Brookfield, H.C. and P. Brown 1963. *Struggle for land: agriculture and group territories among the Chimbu of the New Guinea Highlands*. Melbourne:

Oxford University Press.

Buckley, W. 1967. *Sociology and modern systems theory*. Englewood Cliffs, New Jersey: Prentice Hall.

Cook, S. 1973. Production, ecology and economic anthropology: notes towards an integrated frame of reference. *Soc. Sci. Inform.* **12** (1), 25–52.

Cotgrove, S. 1976. Environmentalism and utopia. *Sociol. Rev. (N.S.)* **24**, 23–42.

Dyson-Hudson, N. 1966. *Karimojong politics*. Oxford: Clarendon Press.

Ellen, R.F. 1978. Problems and progress in the ethnographic analysis of small-scale human ecosystems. *Man (N.S.)* **13** (2), 290–303.

—— 1978a. Ecological perspectives on social behaviour. In *Social organisation and settlement: contributions from anthropology, archaeology and geography, Part I*, D.R. Green, C.C. Haselgrove and M.T. Spriggs (eds.). (B.A.R. International Series [supplementary] 47i). Oxford: British Archaeological Reports.

Evans-Pritchard, E. 1940. *The Nuer*. Oxford: University Press.

Forde, C.D. 1934. *Habitat, economy and society*. London: Methuen.

—— 1953. The cultural map of West Africa, successive adaptations to tropical forests and grasslands. *Trans. N.Y. Acad. Sci.* Ser. II. **15**, 206–219.

—— 1970. Ecology and social structure. *Proceedings of the Royal Anthropological Institute of Great Britain and Ireland for 1970*, 15–29.

Fortes, M. 1976. Cyril Daryll Forde 1902–1973. *Proceedings of the British Academy* **62**, 459–83.

Friedman, J. 1974. Marxism, structuralism and vulgar materialism. *Man (N.S.)* **9** (3), 444–69.

Gellner, E. 1963. Nature and society in social anthropology. *Philosophy of Science* **30**, 236–51.

Glansdorff, P. and I. Prigogine. 1971. *Structure, Stabilité et Fluctuations*. Paris: Masson.

Gluckman, Max (ed.)1964. *Closed systems and open minds: the limits of naivety in social anthropology*. Edinburgh and London: Oliver and Boyd.

Gray, R.E. and P.H. Gulliver 1964. *The family estate in Africa*. London: Routledge and Kegan Paul.

Gulliver, P.H. 1955. *The family herds: a study of two pastoral tribes in East Africa, the Jie and Turkana*. London: Routledge and Kegan Paul.

Hardesty, D.L. 1977. *Ecological anthropology*. New York: John Wiley.

Holling, C.S. 1973. Resilience and stability of ecological systems. *Annual Review of Ecology and Systematics* **4**, 1–23.

Leach, E.R. 1954. *Political systems of highland Burma*. London: Bell.

Lowie, R. 1937. *History of ethnological theory*. New York: Farrar and Rinehart.

McArthur, M. 1977. Nutritional research in Melanesia: a second look at the Tsembaga. In *Subsistence and survival: rural ecology in the Pacific*, T.P. Bayliss-Smith and R.G. Feachem (eds.). London: Academic Press.

MacCormack, C. 1978. The cultural ecology of production: Sherbro coast and hinterland, Sierra Leone. In *Social organisation and settlement: contributions from anthropology, archaeology and geography, Part I*, D.R. Green, C.C. Haselgrove and M.J.T. Spriggs (eds.). (B.A.R. International series [supplementary]

47i). Oxford: British Archaeological Reports.

Mair, L. 1965. *An introduction to social anthropology*. (second edition, 1972). Oxford: Clarendon.

Marett, R.R. 1912. *Anthropology*. London: Thornton Butterworth.

Morton-Williams, P. 1969. The influence of habitat and trade on the polities of Oyo and Ashanti. In *Man in Africa*, P. Kaberry and M. Douglas (eds.). London: Tavistock Publications.

Netting, R. McC. 1971. *The ecological approach in cultural study*. Addison-Wesley Modular Publications 15259.

—— 1974. Agrarian ecology. *Annual Review of Anthropology* 3, 21–56.

Richards, A.I. 1939. *Land, labour and diet in Northern Rhodesia*. London: Oxford University Press.

Royal Anthropological Institute 1951. *Notes and Queries on Anthropology*. London: Routledge and Kegan Paul.

Ruyle, E.E. 1973. Genetic and cultural pools: some suggestions for a unified theory of biocultural evolution. *Human Ecology* 1 (3), 201–16.

Sahlins, M.D. 1969. Economic anthropology and anthropological economics. *Soc. Sci. Inform.* 8, 13–33.

Sahlins, M. 1972. *Stone age economics*. London: Aldine.

Stocking, G.W. Jr. 1968. *Race, culture and evolution*. New York: Free Press.

Thomas, F. 1925. *The environmental basis of society: a study in the history of sociological theory*. New York: Century.

Vayda, A.P. (ed.) 1969. *Environment and cultural behaviour: sociological studies in cultural anthropology*. New York: The Natural History Press.

Vayda, A.P. and B.J. McCay 1977. Problems in the identification of environmental problems. In *Subsistence and Survival: rural ecology in the Pacific*, T.P. Bayliss-Smith and R.G. Feachem (eds.). London: Academic Press.

Vayda, A.P. and R.A. Rappaport 1968. Ecology, cultural and non-cultural. In *Introduction to cultural anthropology: essays in the scope and methods of the science of Man*, James A. Clifton (ed.). Boston: Houghton Mifflin.

Worsley, P. 1956. The kinship system of the Tallensi: a re-evaluation. *J. R. anthrop. Inst.* 86, 37–77.

A TALE OF TWO STATES:
ECOLOGY AND THE POLITICAL ECONOMY OF INEQUALITY ON THE ISLAND OF ROTI[1]

JAMES J. FOX

Marcel Mauss was one of the first anthropologists to introduce a genuine ecological perspective to the study of social organisation. His essay on the *Seasonal Variations of the Eskimo* (1906), written in collaboration with H. Beuchat, contains the essential elements of an ecological model.[2] He defines a specific environmental niche, examines the adaptations of human populations within it, and then, considers both the 'causes' and the 'effects' of these adaptations. In doing this, he clearly recognises the dependence of hunting groups on their animal resources and he charts the consequences of this for patterns of population density, distribution and settlement. Yet, though he insists on the fundamental importance of the material basis of society, Mauss does not assign an extreme determination to this base. Rather he argues explicitly for the subtle interaction of cultural variables at all levels of organisation. Furthermore, instead of relying on a naive synchronism, Mauss turns to historical evidence to distinguish critical features of the system that he is concerned to explicate. Social variation, in time and space, becomes a strategic dimension of his analysis.

I refer to this early essay not merely to point out that the study of ecology has always been a pertinent feature of social anthropology but also to note a difference. Recent developments in our understanding of ecology and the concomitant elaboration of general systems theory now make some of Mauss' groping attempts at analysis seem sadly simplistic. With this added sophistication, we may now be better able to assess the consequences of the interaction of many variables; but we are also more aware of the great complexity of the task we face. Although our present efforts may also eventually be judged as naive, this is no reason to be deterred in our analyses, nor in pushing these analyses in a speculative direction. It is in this spirit that I want to follow Mauss by beginning with a set of ecological variables and proceeding to others of a more sociological nature.

In this paper, I propose to examine some of the material bases for social variation on the island of Roti in eastern Indonesia (see Figure 1), and I want to consider this variation in historical perspective. To do this, I shall focus on two small states on the island which show significant differences in their social and political organisation. I intend to consider how ecological, sociological and cultural variables have interacted in their divergent development.

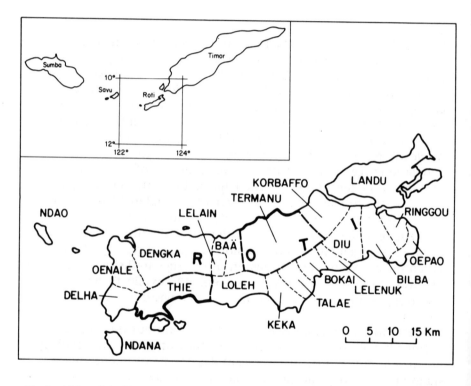

Fig. 1. A Map of the Island of Roti showing its various small states or domains: Termanu and Thie are indicated in darker outline.

The background to this paper is contained in my book, *Harvest of the Palm: Ecological Change in Eastern Indonesia* (1977a). Initially, I shall have to summarise, as briefly as I can, some of the analyses in this book. My intention, however, is not to repeat what I have already written but to provide a minimal context within which to develop one of the arguments in the book a step further.

The Ecology of Palm Utilisation: A Basis for Economic and Social Variation

Roti, together with the islands of Savu, Sumba and Timor, forms part of the outer arc of the Lesser Sundas. In this region of Indonesia, a variety of factors have combined to create a critical situation of increasing aridity. A long and windy dry season alternates with a brief period of limited, irregular rain; the shifting agriculture of an expanding population has produced extensive deforestation and con- tributed to extreme erosion and soil impoverishment; and on the large islands of Timor and Sumba, severe overgrazing has resulted from commercial attempts to raise cattle (and/or horses) for export while the rapid spread of recently introduced shrub-weeds has further complicated the situation by placing new constraints on native agriculture. For twenty years now, Timor — in particular — has been cited by geographers as a classic example of a 'problem' case in human ecology (Ormeling 1956).

Within this region, the deteriorating swidden economies of west Timor and east Sumba show a marked contrast to the productive palm economies found on the small islands of Roti and Savu. These islands, the southernmost of Indonesia, are subject to even worse climatic conditions than either Timor or Sumba, and yet they support populations several times those of the larger islands without seasonal or sustained periods of hunger. Roughly stated, the Atoni of Timor with their swidden economy have experienced serious problems when and where their density has gone beyond 30 to 35 persons per km^2 whereas some areas of Roti and Savu have reached population densities of over 100 persons per km^2 without showing signs of similar problems. (For a more detailed statement of these varying densities for the various islands since 1930, see Fox 1977a: 17–23.) Because of their somewhat unusual nature, the economies of these small islands deserve special investigation.

The palm economies of Roti and Savu involve a combination of specialisation and diversification. The lontar, a species of the fan-palm *Borassus* (along with a species of *Corypha* found mainly on Roti) provides a remarkable array of products ranging from construction materials (timber, fencing and thatch) and household articles (baskets, buckets, mats) to objects of wear (hats, belts, workclothes and even such things as disposable sandals). More importantly, the palm provides a major source of food for human and animal sub- sistence. The juice which the lontar produces in abundance is either drunk fresh or cooked to a syrup for storage; it is a staple of the daily diet.

A secure dependence on palms, whose primary exploitation occurs during an otherwise slack period of the year, allows a considerable diversification of other pursuits: a combination of fishing, livestock-raising and gardening. Each of these pursuits functions as a highly variable subsystem within the economy. All Rotinese and Savunese, for example, raise the same kinds of animals — waterbuffalo, pigs, goats and sheep; but it can be shown that the combination of these animals — their ratio to one another in any total configuration of livestock — varies considerably. Local conditions make herding strategies significantly different while livestock-keeping is also related to palm utilisation. Since surplus syrup is fed to pigs, thus converting excess carbohydrates to usable protein, pig-keeping increases in direct proportion to the intensification of palm utilisation.

On these islands, semi-permanent gardening occurs as opposed to shifting agriculture, and this gardening is not based on any single crop as has increasingly become the case among the swidden populations on neighbouring islands. Most gardening involves multi-cropping. The predominance of any one crop — either maize, millet, sorghum or green grams (mung beans) — varies according to different local factors.

On Roti, the cultivation of rice introduces another important factor in this situation. Wherever there exists the possibility of growing rice — by means of permanent irrigation, by the diversion of rivers or streams, or by means of rain-catchment — this opportunity is invariably seized upon. Thus Roti presents an odd patchwork pattern consisting of a few well-irrigated areas of wet-rice combined with innumerable small and scattered fields where rice may be grown with reasonable success in favourable years. Variation over time is critical to the overall functioning of the agricultural sector of the Rotinese economy. Because the rains are highly irregular, only certain sawah areas produce rice with any degree of consistency. Yet every few years, the island as a whole (or large areas on it) can produce an abundant rice harvest; in other years, the Rotinese agricultural strategy aims at a more modest success based on some mix of garden crops.

In short, this kind of palm-centered economy is a flexible and productive system. On Roti, it has allowed the development of considerable social variation among local populations of widely-differing densities.

Modes of Explaining the Rotinese and Savunese Expansion

The Rotinese and Savunese, however, are not confined to their small islands. They are to be found both on Timor and Sumba where they occupy areas of palm savanna and have managed to transfer virtually intact the whole of their distinctive economies. As potential competitors for the same kind of niche, the Rotinese and Savunese appear to have avoided competition by moving mainly to separate islands: the Rotinese to Timor, the Savunese to Sumba. If one adds to this the fact that present evidence indicates that the *Borassus* and *Corypha* palms have followed in the wake of swidden climax, one might be tempted to see in this a clear and simple case of ecological succession: palm-tappers replacing swidden cultivators as they themselves alter their environment. Where once there were Timorese or Sumbanese cultivators, there are now Rotinese or Savunese palm-tappers.

Historical records going back over three hundred years make it clear, however, that this has not been so simple a process. The long involvement of Europeans – Portuguese and Dutch – in local affairs on these islands and the successive policies of the Dutch East India Company and the later colonial government had a great deal to do with 'channeling' these populations to the 'niches' that they now occupy. On its own, an ecological explanation is inadequate to explain the present situation.

Moreover the Rotinese and Savunese have very different languages and cultures and they responded to European presence in notably different ways. The historical evidence shows that the separate migrations of the Rotinese and Savunese to Timor or Sumba took place at different rates, at different times and for different internal reasons (Fox 1977a: 127–76). Thus it would seem that in dispelling a simplistic view of a complex social process, historical particularities dissolve the possibility of discovering any pattern to this process. But this is precisely what I do not wish to contend. My object in this paper, as in my book, is to attempt to integrate historical and ecological data in a more comprehensive analysis. The rest of this paper will be devoted to this end.

An Outline of the Major Local Differences on the Island of Roti

If one divides the island of Roti crudely into two more or less equal halves, one may begin to glimpse something of a pattern. On Roti, water resources, for example, are unevenly distributed. The majority of reliable sources for wet rice irrigation are found in the eastern half of the island and many of them occur at a slight elevation along the island's central spine thus creating a reasonable gradient for directing water toward the coast. As a result, east Roti, in comparison with west Roti, is relatively rich in sawah. By contrast, west Roti, which is drier, is noted for its more intensive palm utilisation.

These differences are reflected in livestock-raising strategies. West Roti shows the heaviest investment in pigs whereas east Roti has the heaviest investment in waterbuffalo. Because they are fed on lontar-products, the raising of pigs is closely related to palm utilisation. But the keeping of waterbuffalo is closely related to wet rice cultivation since the Rotinese rely on the waterbuffalo to tread flooded fields into a mud bed before rice can be planted (Fox 1977a: 38—47).

One can go further in this contrast by introducing another point of comparison. Except for one or two small areas, conditions on Savu are very much like those on west Roti. If anything, Savu is even more reliant on the lontar than west Roti. Using other criteria of assessment, Savu clusters closely with the domains of west Roti: 1) in its total livestock-keeping pattern, 2) in its emphasis on pig-raising and 3) in the correlation of pigs to the human population (*ibid.*: 45—7). More significantly still, the demographic history of Savu for over a hundred years has more or less paralleled that of west Roti. It has definitely not paralleled that of east Roti (*ibid.*: 169—70). The effects of emigration from east Roti account, in large part, for the overall difference in the demographic histories of the two islands.

The fact is that all areas of Roti and Savu have contributed emigrants to the larger islands but not in the same proportion. This is strikingly clear in comparing east and west Roti. The population density of east Roti has declined since 1861 whereas the population density of west Roti (and Savu) has gone up significantly. A more detailed analysis of these figures shows a seeming paradox that cannot be explained by a population pressure model for emigration. Certain areas of relatively low density have had a high level of emigration whereas other areas of relatively high density have had a proportionally low level of emigration (*ibid.*: 156—9). Many factors are involved here. Since, however, Savu and west Roti show roughly the same pattern, I want to examine the factors that make east Roti so different.

Dutch Involvement in the Population Reduction of East Roti

Dutch involvement in local affairs must be recognised as one of the contributing factors to the relatively lower population density of east Roti. It is necessary, however, to distinguish a single radical intervention in the eighteenth century from more consistent and recurrent actions in the nineteenth century for which there was a general policy.

Landu, comprising twenty-nine percent of the land area of east Roti, is distinguished by the fact that it has the lowest population density of any small state on Roti (16 persons per km²). In many other respects, Landu is economically unlike the rest of the island. But to look for an ecological explanation of these facts would be misconceived. One historical event overshadows all other factors in explaining present conditions.

Landu was among the first domains, in 1653, to swear allegiance to the Dutch East India Company. To judge from the levies of men that Landu agreed to supply to the Company, it was, at the time, one of the most populous areas of Roti. By 1750, its population was estimated at 4,000 which would indicate a density of at least 23 persons per km². But in 1756 Commissaris J.A. Paravicini decided on a simple solution to a protracted dispute over succession within the domain. He sent 50 armed Europeans, 50 Balinese and all the local native manpower that he could muster in the ship, *De Vrijheid, The Freedom*, 'to raze and ruin [Landu] once and for all, to take everyone as a prisoner of war, even the supporters of the regent . . . and to allow not a single Landunese to hide there, nor even to dwell on the island of Rottij' (Chrijs 1872: 222). This order was carried out to the full. Those, who did not die in the first onslaught or who did not starve to death among the group who attempted to hold out on one hill-top, were hunted, captured and declared slaves. A large number were divided among those who took part in the raid, including Company officers and soldiers. According to one report, the Commissaris received 300 slaves as his personal share of the plunder and 1,145 men, women and children were sent to Batavia for the profit of the Company.[3] In one fell stroke, Landu became a no man's land and, in time, something of a nature preserve.

Nothing in the internal workings of any model of Rotinese society would predict this sudden, selective and totally devastating change that occurred in Landu. But knowledge of this event can serve to highlight other more general features of the colonial situation. The question is why Landu was not rapidly repopulated in the nineteenth century. Again, the answer to this is not to be found in an ecological explanation. It lies in the sphere of politics and specifically in the political control by local rulers of their subjects — a control which the Dutch sanctioned and enforced.

From the earliest records of the seventeenth century, the Rotinese were consistently represented as a fractious and disputatious population. The Dutch were quickly drawn into these disputes. Most of them, including the dispute that ended in the depopulation of Landu, were dynastic disputes. The initial response of the Dutch was to take sides with certain rulers and to devastate their rivals. This gave way to a more moderating policy in which the Dutch settled disputes by recognising and conferring titles on more and more local rulers, a policy which led to extreme political fragmentation. By the end of the eighteenth century, the tiny island of Roti was formally divided into eighteen small states, each of whose ruler had the title of *radja* or *regent*. Though by no means the most populous of islands, Roti held the peculiar distinction of having the highest density of rajas of any island in the Dutch East Indies (Fox 1977a: 92–112).

In the nineteenth century, the Dutch adopted an entirely different attitude toward dynastic disputes. At the time, they needed armed men to settle in the partially depopulated buffer zone between them and the hostile Timorese whom they could not control. They turned to Roti for this manpower but they used this need to solve local dynastic quarrels. Here there occurred a coincidence of interests between local rulers and the Dutch authorities. Certain rulers consolidated their position by cooperating in the exclusion of their dissidents, opponents, and rivals (Fox 1971a: 65–6; 1977a: 136–42). Not all of this involved force; some of the emigration in the later half of the nineteenth century was more spontaneous and was partially motivated by population pressure (Fox 1977a: 127–59). But the pattern seems to have remained much the same. Research on Rotinese villages in Timor suggests that most were founded by high nobles or members of royal lines who led pioneer groups from Roti to Timor.[4]

All of this leads to a more specific phrasing of the problem at hand: what internal factors led certain states to contribute a selectively higher proportion of their population to the emigration to Timor? To try and answer this question, I want to consider cultural categories common to the entire island of Roti – those that have to do with descent, alliance and exchange, all of which are fundamental to matters of social reproduction.

Cultural Categories of Social Reproduction and Exchange

The island of Roti, with its population of approximately 76,000 persons, possesses a culture which is characterised by variation among the eighteen small states (or domains) that until recently formed its recognised political units. In my view, there is a unity, though not a uniformity, to this culture. The fact that all dialects spoken on the island are mutually intelligible is some justification for this view. This alone is insufficient. I would contend that many differences on which the Rotinese pride themselves (Fox 1974; 1977a: 79–85) are elaborations, within the political context of a former colonial situation, of a single system of basic cultural categories. This is not, in any way, intended to underestimate these differences. Minor alterations in categorical relations can have systemic consequences in social organisation. Categories can remain consistently intelligible as their usage is altered.

On Roti, descent groups are constituted by a succession of personal names. A predominant and presumptive male bias in the succession of names,[5] coupled with a similar bias to the inheritance of property and of ritual positions, gives these groups an agnatic structure. Everywhere on Roti, the highest level descent group is called a *leo*. There are more than 200 named *leo* on Roti and I shall refer to these groups as 'clans'. Since political, rather than demographic factors, have been uppermost in the formation of these clans, they vary enormously in size and in internal structure. Lesser level segmentation may occur within clans and to describe this the Rotinese share a common terminology (Fox 1971a: 41–2; 1971b: 224–5). The most important of these terms is *-teik* to which a proper name is prefixed. It is this term that I shall translate as 'lineage'. Although titles, prerogatives, and certain ritual rights pertain to clans or lineages, all property is held by 'households' (*uma*) of limited size whose development is directly related to the life-cycle of its members.

On Roti, male relations within groups contrast with female-linked affiliations among them.[6] Marriage is the only means of creating these affiliations.[7] Every marriage establishes a specific alliance marked by ritual performances and obligatory exchanges. Bridewealth payments initiate this alliance and mortuary payments conclude it. These payments are the price of a relationship. They are given in return for the life-giving powers which this relationship provides. Wife-givers are thus 'life-givers' and must be continuously compensated for their services.

Every individual recognises a specific chain of life-givers. Sociologically, this is a line of maternal affiliation. The length of this line (or the number of generations over which it is formally recognised by performance and exchange) varies somewhat among the domains of Roti. Everywhere, however, this line comprises three generations and nowhere is there an obligation that it be renewed. Theoretically, every marriage creates a new alliance and forges a new link in a chain of relations. The concept of 'prescription' in such matters is irrelevant to Roti.

Each household (*uma*) negotiates the marriages of its members and, from its resources, covers the major share of all exchange payments. Since these groups are the least determinate units in the society, marriage is generally spoken of as linking lineages, clans or even larger groupings. A minimal requirement is that marriages should adhere to whatever rules regulate relations among members of these higher order groupings. The markedly different structure of Rotinese descent groups, however, makes alliance relations among them equally various. What all of these alliances share is a similar imagery.

Marriage is seen as a union of male and female principles which are manifest, for example, in the sun and moon. Marriage may also be seen as the delayed reunion of a brother and sister pair via their descendants. In either case, once established, the imagery of the relationship is pre-eminently botanical. The mother's brother as wife-giver 'plants' his sister's children and, throughout life, sustains and nurtures these plants.[8] In Rotinese, this botanic idiom also involves a tangle of other metaphors. In sexual intercourse, semen unites with female blood to create bone and flesh. Bone is regarded as durable and flesh as vulnerable. Thus the slightest physical injury or even the accidental letting of blood requires the payment of compensation to the mother's brother. A final reckoning occurs at the mortuary ceremony where various persons in the line of maternal affiliation cooperate to conduct the death rituals and are all fittingly rewarded. What endures, after death, is a named spirit that takes its place of succession within the descent group and bones which, though committed to the grave, are described as endowed with a symbolic continuity.[9]

The categorisation of required objects of exchange also conforms to a recognised pattern. Wife-givers are given 'male' objects such as swords, spears, gold, herd animals (waterbuffalo, goats and sheep), and raw meat;[10] and they reciprocate with 'female' gifts of cloth, ancient glass beads, pigs and cooked food. The same classification holds for marriage and mortuary exchanges since these form the beginning and end of a single exchange cycle. Some attempt is made to match objects of exchange but not for the purpose of achieving equal value. All exchanges are weighted in favour of wife-givers. This only increases over the duration of the cycle, especially since the wife-givers' ritual services are a part of the transaction and must be rewarded. On Roti, beautiful symbolic worlds are built on a firm material foundation.

The Political Elaboration of Social Differences among Rotinese

Having described certain cultural similarities, I want to concentrate on the differences among Rotinese. It is these differences that the Rotinese exalt and exaggerate to distinguish themselves. Two factors make Roti a particularly interesting case study. The first has to do with the divisive political fragmentation of the island which, as it happens, can be well documented; the second has to do with the Rotinese historical understanding of this political process.

The earliest Dutch documents report a number of petty feuding domains on Roti. In 1662, the Dutch began by signing treaties with a few of the more powerful of these domains. By the end of the eighteenth century, however, they had given full formal recognition to eighteen separate states. Each of these domains (*nusak*) was organised around a dynasty whose court, attended by titled representatives of all constituent clans, had the right to settle disputes by local customary law. Existent dialect variation was fostered to provide the basis for a native political dogma that each domain has its own 'language' to correspond with its distinctive court law.[11] Each domain is thus conceived of as distinct. The number of clans, their arrangement, the allocation of titles, prerogatives and ritual privileges — all coalesce to define a particular domain.

History generates diversity, and, in the Rotinese view of things, history provides the key to understanding political affairs. The configuration of present day relations is the result of specific events that have occurred in the past.[12] The separate genealogy of each domain's ruling dynasty provides the chronology for these events.[13] These genealogies are immensely long and complicated. The tales that accompany them document political developments which, to the Rotinese, began long before the arrival of Europeans.

Marriage, under these circumstances, is more than just a symbolic enactment of an archetypal pattern. The exchanges which marriages initiate are part of each domain's political economy and, to the Rotinese, a domain is not considered to have a single, simple 'mythological' foundation. It is regarded as having developed over time through the acts of the ancestors. Its development is part of a continuing history.

Termanu and Thie: Systemic Differences between Two States on Roti

Here I propose to examine two domains — Termanu and Thie — which exemplify various opposing tendencies in their systems of marriage and exchange. My first purpose is to consider the way in which similar cultural categories are given different quantitative valuation and are, thus, used to support political structures of relative hierarchy or equality. My second purpose is to link these differences to other variables that I have already discussed.

Termanu, in east Roti, is, for example, renowned for its wet rice irrigation; Thie for its reliance on the lontar palm. Consistent with this, Termanu shows a high investment in waterbuffalo and a relatively lower investment in pigs; Thie shows exactly the reverse (Fox 1977a: 47). Thie with a population of over 130 persons per km² is well above the average density for west Roti while Termanu with only 32 persons per km² is well below the average even for east Roti. Many other factors are also involved in this contrast – not the least of which is the conscious opposition to one another that these two domains have maintained since at least the seventeenth century. Here, however, I want to outline some significant social and political differences within these domains. My contention shall be that all of these differences form part of a larger pattern.

Termanu dominates a large area of east Roti. At the time of the arrival of the first Europeans, it seems to have been on the verge of dominating the entire island. The historical narratives of the domain acclaim the conquests of Termanu's ruling dynasty whose expansion began with secure control of some of Roti's richest sources of irrigation in the very centre of the island. By contrast, Thie's narratives describe their rulers' retreat to a peripheral point in the southwest of the island. These narratives even admit that Thie was, for a time, a tributary to Termanu.

Termanu was formally recognised by the Dutch in 1662 and its rulers were accorded singular honours. Initially the Dutch relied on Termanu in their dealings with other rulers on the island and prestige goods filtered through Termanu to other areas. At first Termanu seems to have used its position to stifle opposition within its territories. Thus, for example, Thie suffered severe reprisal in 1681 for purportedly rebelling against the Company. But eventually, in 1690, Thie was officially recognised, though it is not clear whether this altogether altered its tributary status in regard to Termanu.

Only after 1725 did Thie's rulers resort to the radical innovations that secured their autonomy. A ruler of Thie converted to Christianity and, as head of a 'Christian state', formed an alliance with the rulers of several other small domains who followed Thie's initiative in converting and opposing the pagan rulers of Termanu. As opposition increased in the eighteenth century, Termanu suffered the loss of much of its former territory. Termanu was the chief victim of the Dutch policy that led to the political fragmentation of Roti.

In the nineteenth century, the rulers of Termanu adopted a different strategy and, in cooperation with the Dutch, expelled whole villages of dissidents to pioneer settlements on Timor. This, in turn, facilitated the further emigration of members of Termanu in the late nineteenth and early twentieth century. By contrast, Thie — hemmed in on the south coast — contributed proportionally less to the overall emigration of Rotinese to Timor.

This history is reflected in the population figures for the two domains. Termanu still controls almost twice as much territory as Thie but ever since the population expulsions in the first quarter of the last century, Termanu's population has never equalled that of Thie. At present, Termanu has less than half the population of Thie and only one-fourth of its population density.

The year 1831 presents a useful historical base line for comparing the two domains. At that time, both Thie and Termanu were made up of ten clans — a royal clan plus nine other clans whose representatives at court were known as the *Manesio*, 'Lords of Nine'. If one can accept the census figures of this period, it is apparent that both domains have shown a population increase of roughly the same magnitude (3.5 x for Thie; 3.0 x for Termanu).[14] In response to this, the number of clans (*leo*) in Thie has increased from 10 to 26. In Termanu, however, no new clans have come into existence. Termanu's rulers endeavoured to maintain their domain's ideal structure of nine clans. But since the court has adamantly refused to recognise the formation of new clans, some of Termanu's original clans have become highly segmented. There are now over 40 named lineages (*-teik*) within these clans. Whereas, in Thie, the clan (*leo*) forms the exogamous unit of the domain, in Termanu the lineage (*-teik*) has become the exogamous unit in all large clans.[15] Most clans in Termanu permit endogamy.

The contrast between these domains goes much further. The only lineages that are said to exist in Thie are those in the royal clan. They are relatively shallow and of nominal significance, serving mainly to distinguish various royal descent lines. In terms of marriage, the royal clan is exogamous.

In Termanu, the greatest proliferation of lineages has occurred in the royal clan. They are ranked and stratified and each is its own marriage unit. As a consequence, the royal clan in Termanu is markedly endogamous.[16]

The greatest difference between these domains lies in their overall marriage structure. Thie possesses a moiety system which is regarded as the hallmark of its political identity.[17] Its clans are divided into two groups, Sabarai and Taratu, which intermarry with one another. The clans of Sabarai are under the jurisdiction of the ruler or 'male lord' (*manek*), while the clans of Taratu are assigned to the secondary, executive lord, known in Rotinese, as the 'female lord' (*fetor* or *mane-feto*). At a general level, marriage in Thie is supposedly a symbolic union of 'male' and 'female' but, as anyone in Thie will admit, the system does not actually work at this level. Five clans in Sabarai with ancient ritual rights over the earth form a group that may marry with Taratu or with other clans in Sabarai.[18] Thus, instead of two, Thie consists of three large intermarrying groups. Given the existence of court-imposed sanctions for marriage violations, these three groups do marry according to rule. In a survey of marriages in the adjoining village areas of Oe Handi and Meo Ain in Thie, 95 per cent of all marriages conformed to this proscriptive pattern of marriage.[19]

A significant feature of this system is that all clans are theoretically equal and hence the bridewealth paid for any woman in Thie is supposed to be the same. Since before the middle of the nineteenth century, the rulers of Thie have taken an active part in maintaining a schedule of equivalent bridewealth for all their subjects and are reported to have imposed heavy fines on those who tried to insist on bridewealth in excess of this standard.[20]

The rulers of Termanu did just the opposite of the rulers of Thie. They promulgated a schedule (or scale) of differential bridewealth based upon the status of the wife-giving group.[21] And instead of intervening to maintain a fixed maximum, these rulers, as the most aristocratic members of the highest lineage of the highest clan, simply presided at the top of a competitive system whose escalation (and inflation) was to their advantage. This produced an endogamous spiral of marriage relations among those clans with 'status' lineages.[22] The hierarchy of lineages within these clans replicated the hierarchy of clans within the domain.

Two further features highlight the structure of relations in Thie and Termanu. In Thie, as a rule, bridewealth (or a significant portion of it) should be paid before marriage. In Termanu, this is not the case. In fact, it is possible, as the saying goes in Termanu, for a 'daughter to pay the bridewealth for the mother'. In other words, the unpaid portion of a woman's bridewealth may be taken from the bridewealth paid for her daughter simply by altering the proportion due to the girl's mother's brother.[23]

In both Thie and Termanu, individuals go through the same cycle of ceremonies which require similar symbolic exchanges. In Thie, however, persons pay their way as they proceed through these rituals

TABLE I

Schematic Outline of Structural Differences between Termanu and Thie

Termanu, East Roti	Thie, West Roti
Central Domain, North Coast	Peripheral Domain, South Coast
Prominence of Wet Rice Cultivation	Intensification of Palm Utilisation
High Proportion of Waterbuffalo	High Proportion of Pigs
Relatively Low Population Density: 32.5 persons per km² (1961)	Relatively High Population Density: 130.3 persons per km² (1961)
High Emigration Relative to Population	Low Emigration Relative to Population
No Change in the Formal Number of Clans (*leo*) since 1831	Increase in the Number of Clans (*leo*) from 10 to 26 since 1831
Proliferation of Lineages (*-teik*) within Clans	Virtually no Development of Lineages
Ranking of Clans and of Lineages within Clans	Equality of all Clans
Genealogical Links chiefly among Clans with Status Lineages	Putative Genealogical Links Embracing all Clans
High Degree of Endogamy among Clans with Status Lineages	Strict Exogamy of all Clans
Competition among Status Lineages: Relatively Little Court Interference in Marriage Arrangements	Marriage Regulation via Moiety System: Court Supervision and Interference
Differential Bridewealth based on the Status of the Bride-giver	Fixed and Equal Bridewealth for all Women
Delayed Payments in Ritual Exchanges	Direct and Immediate Reciprocity in Exchanges
Greater Recognition and Dependence on Line of Maternal Affiliation	Lesser Recognition of Line of Maternal Affiliation

whereas in Termanu, the emphasis is on delayed payment which creates a sense of debt and dependence over generations. This is particularly evident in mortuary ceremonies which are a more important 'point of reckoning' in Termanu than they are in Thie.

A concomitant of this is the greater formal recognition given the line of maternal affiliation in Termanu. At mortuary ceremonies, both the mother's brother and the mother's morther's brother are paid for their services. In Thie, only the mother's brother must be given gifts.[24]

A Tale of Two States: A Tentative Interpretation

Here I would like to try to pull together the various strands of this paper. I propose this as my tale of two states — a tentative, possible outline of a complex developmental process. Since this process began before the arrival of Europeans in this area, we can only speculate on its early foundation, but at least Dutch records, beginning around 1650, provide indications on the course of this development.

We must begin with the fact that in the middle of the seventeenth century, Termanu was the most prominent 'domain' on the island. Dutch records indicate this clearly and most Rotinese tales corroborate it. Termanu's historical narratives present the domain's rise to power as a process of conquest and incorporation by a succession of war leaders (*palani*) of what is now the royal clan. Termanu's commoner clans are said to have originated either from those who were con-quered or those who were allied to Termanu's rulers. Virtually all the disputes that these narratives recount concern sources of water which eventually became the means of irrigation for rice. These conquests are not, however, attributed to force of arms but rather to cunning and deception, a feature that has gained Termanu its proud sobriquet, *Kekedi Pada*, 'Perfidious Pada' (Pada being the more common Rotinese name for Termanu). One suspects that at this time Termanu presided over a loosely structured association of local feuding groups. This association perhaps ought not to be called a 'state'. This is essentially what the narratives assert. The change from war leader (*palani*) to lord or ruler (*manek*) is attributed to the Europeans. Furthermore the Council of Rajas that was organised and convened in Kupang by officers of the Dutch East India Company bears too much similarity to the organisation of Rotinese courts for this influence to be discounted.

Termanu's early prominence was undoubtedly based on its control of valuable areas of wet rice irrigation, and the raiding that Termanu continued to carry out into the eighteenth century was timed and directed to obtain both rice and heads. Here certain cultural values play their part. Roti is not particularly well-suited for the growing of rice as compared to other crops. Yet rice is the necessary require-ment for all feasts. Lontar products were already abundant at this time. Rumphius, writing in the late seventeenth century, reports that the Rotinese 'hang [buckets of lontar juice] in their houses and anyone who comes there may freely drink of this and so satisfy his thirst, even before he has greeted anyone, this being an act of polite-ness and custom in that land' (Fox 1977a: 78). But both because of its abundance and because of certain cultural values, lontar syrup never rivalled rice. A surplus of rice is not the same as a surplus of syrup. Rice is a scarce commodity, a prestige food, and a means of ceremonial payment. Lontar syrup is none of these.

Another factor in Termanu's prominence was its control not just of water but of what the Rotinese call *naü-oe*, 'grass and water'. Directly to the east of Termanu's central *sawah* areas are the best grazing lands on the island. Termanu's expansion was directed eastward as well as westward.

The biggest boost to Termanu's position, however, occurred with the arrival of the Dutch. Termanu was among the first to sign a contract with the Company. The Company was allowed to establish a small post at *Kota Leleuk* on Termanu's north coast. As a result, the Dutch tended to rely on Termanu's rulers as their spokesmen and interpreters. In 1679, they brought the young ruler of Termanu, Pelo Kila, to Kupang to learn Malay and until his death in 1718, Pelo Kila acted as their chief advisor on the island. Elite goods, including weapons and cloth, seem to have passed via Termanu to the rulers of the other domains. Other trade followed this same route. Better access to weapons may have aided Termanu but there is not much evidence of this. But the evidence of the importance of cloth for Termanu is unmistakable. To this day, all nobles on Roti are identified by their right to wear specific motifs on their *ikat* cloths. Termanu's design system of noble cloths is the most elaborate on the island, and the majority of the motifs in this system can be traced back to the *patola* cloths provided by the Company (Fox 1977b).

The segmentation that eventually gave rise to the lineages of Termanu's royal clan derives from the eighteenth century. The foundation of Termanu's elaborate status system was established at a time when the domain was at the height of its power. We can be fairly certain of this because most of the ancestors who are named as lineage founders can be dated with reasonable accuracy in the Dutch records.

A potential for status differentiation is inherent in any system that composes lineages on an elder/younger sibling distinction. Segmentation is always segmentation with a difference. But something more than this is needed for the creation of status lineages: an expanding resource base that allocates wealth and prestige according to this distinction. Termanu, I would contend, had both of these ingredients between roughly 1660 and 1760.[25]

Such a system is stable provided that it can expand. After about 1735, however, Termanu's position began to be challenged. Its rulers, who remained adamant traditionalists in the face of an encroaching Christianity, were unable to maintain their unique access to the Dutch. Several crucial Dutch decisions curbed Termanu's rulers. They lost their power to raid and to obtain tribute; and after 1750, Termanu became a victim of increasing political fragmentation. Eventually in the early nineteenth century, the Dutch established a new administrative post in the domain of Baa and transferred all their activities to this site. Termanu's power was not merely checked; it was actually diminished.

Termanu's rulers faced conditions in the eighteenth century unlike those of the previous century. No other domain possessed — at least in embryo — as elaborate a status system within its ruling ranks and no other domain[26] did precisely what Termanu did, namely, refuse to alter its basic structure of nine clans. This refusal introduced a new form of instability to the system which was, in fact, only relieved by the 'expulsion' of large segments of the population. My proposition is that a status lineage of this sort is essentially expansionist but when this expansion is curbed, it must 'export' population to survive.

Termanu's status system is both open and exclusive. Wealth, in the form of waterbuffalo, moves up as some women marry down. But there are no restrictions on who may marry women of high status. Nobility is inherited through the male line; status, visible — for example — in the right to wear specific noble motifs, is transmitted through women. This means that nobles must marry nobles of equal or higher rank to retard the process of inherently diminishing rank. Wealthy commoners can gain status for a generation by marrying noble women and many do marry into the lesser ranks of nobility. By this same token, emigrants who have gained wealth abroad may return to marry high-born women. The system is also open to a form of clientage. Thus, for example, when the Dutch transferred to Baä, a Chinese merchant appointed a Savunese noble as his representative in Termanu. This man, Doko Bire, was extremely successful and he and his descendants translated this success, by a recurrent pattern of marriages at the highest level, into a unique position within the royal clan. In time, the highest of these descendants became the *wakil* or deputy of the ruler himself (Fox 1968: 242–55). Similarly, Termanu used its prestige within Roti to give women to the rulers of other domains. Although he never explained matters in detail, the ruler of Termanu to whom I was court historian claimed a relation by marriage to all but one of the eighteen traditional ruling lines.

Termanu's status system is, however, essentially twofold. Commoners may marry among themselves at a fraction of the cost of someone in Thie. Although individual commoners may opt to pursue a limited status, all but one of the commoner clans in Termanu are excluded from the system. In the nineteenth century, pressure was heaviest on the noble class, the class of the *mane-ana*. They were excluded from office at court, since the court was a collection of representatives of all clans and they faced the prospect of diminishing rank. The polygyny of the noble class which was formerly part of this same system only increased its gradient. From all of the evidence that I have been able to gather, it was the *mane-ana*, roughly in proportion to their rank, who were the main source of 'unrest' in Termanu. Many of these *mane-ana*, when they failed to achieve their ends within the domain, gathered a following and led them to Timor where they could become mini-lords of new lands. The educational system in which the Rotinese had an enormous advance over other people in the area operated in the same way but on a more individual basis.

There is a special linkage in Termanu's exchange that is crucial to its operation — the relationship between rice and waterbuffalo. Buffalo are more than a mode of exchange; they are a means of production. An increasing concentration of buffalo can be translated into an effective expansion in rice production. The accumulation of waterbuffalo and wet rice fields in high noble hands is clearly indicated in Termanu, and it is my suspicion (which I am attempting to investigate) that the greatest expansion of Termanu's traditional 'rice bowls' (*Pinga Peto ma Soë Lela*) occurred in conjunction with its spiralling status pursuits in the nineteenth century.

The tale of Thie makes a valuable contrast at this point. As a beleaguered small domain in the sixteenth and seventeenth centuries, it never developed the internal status differentiation of Termanu and in the nineteenth century, its rulers opted to increase the number of clans and to enforce an equality among them. Nor did Thie have the same relative possibilities of linking waterbuffalo to rice production. A wealth in pigs is not the equivalent of wealth in waterbuffalo, 1) because these animals are not raised as herds on open grazing land but on the surplus syrup of individual households and 2) because they are only allowed to pass in the same direction as women and are, thus, superfluous to a system based on the status of women. It seems to me that Thie may never have had enough waterbuffalo to operate its marriage system as it should properly have worked, especially if some of these animals were used in feasting. This might explain the notable tendency in Thie to pay bridewealth in gold and/or to demand immediate payment of waterbuffalo.

To conclude this tale of two states, I would like to point out one last feature of Termanu's state structure. A crucial feature of any system based on the superior status of wife-givers revolves round the problem of who is to be wife-giver to the state's highest wife-giver. This problem can always be solved by bringing women in from outside but an internal solution deserves special note. In Termanu, by long tradition justified in a legendary charter, the royal line of Fola-teik in Masa-huk, takes wives from one lineage, Tulle-teik, of the commoner clan, Kiu-Kanak with whom it divides water rights over the rice fields from which it originally expanded.

Rotinese historical tales end with a formula that states that they are true. I shall be less assertive and merely state this is about as far as I can go for the moment in my continuing research on the island of Roti.

Footnotes

1. The research on which this paper is based was originally supported by a Public Health Service fellowship (MH-23, 148) and grant (MH-10, 161) from the National Institute of Mental Health and was conducted in 1965—66 in Indonesia under the auspices of the *Lembaga Ilmu Pengetahuan Indonesia.* The continuation of this research was again supported by a NIMH grant (MH-20, 659) and carried out in 1972—73 under the joint sponsorship of LIPI and the University Nusa Cendana in Kupang, Timor. This paper incorporates material from an earlier paper of mine, 'Equality and Hierarchy in Rotinese Symbolic Exchange' which I read at a Symposium organised by Shephard Forman and Clark Cunningham at the American Anthropological Meetings in Houston, Texas in 1977. I want to express my thanks to the members of this symposium including Edward M. Bruner, the chief discussant; to C.J.G. Holtzappel of Leiden University; and to attendant members at the ASA Conference in Cambridge, in particular Roy Ellen and Sir Raymond Firth, all of whom offered me comments on earlier versions of this paper. I also wish to thank the Board and members of the staff of the Netherlands Institute of Advanced Study in Wassenaar for the opportunity to pursue my researches and to write this paper.
2. I have recently completed a translation of this classic work by Mauss which will be published by Routledge and Kegan Paul in the near future.
3. The 'Landu Affair' was one of the most complicated disputes that occurred in a century of complicated disputes. I do not claim to understand the entire basis of the dispute. But since the Commissaris' actions were seriously questioned at the time and were cited in the nineteenth century colonial literature as an example of gross misconduct, there exist some publications on the subject. See Chrijs (1872) and Leupe (1877). The unpublished literature is considerably larger.
4. Hence these villages replicate much of the structure of a Rotinese domain and have remained more or less intact as such for over a hundred years.
5. It ought to be noted that the Rotinese naming system could work equally well through males or females or through some combination of male and

female names. As such, the system itself is essentially neutral and could, in time, be adapted in various ways. In the Rotinese case, we are dealing with one possible interpretation of a more general system. The fact that a woman's name can, on occasion, provide a link in a chain of succession only serves to underline this crucial feature.

6. To be more precise, relations within groups are categorised in terms of elder/ younger sibling of the same sex whereas relations among groups are spoken of in terms of siblings of the same sex. As with naming, this system is neutral but is given an exclusively male interpretation on Roti. When elder/ younger terms are used to describe internal group relations, the reference is to males. And for opposite sex siblings, it is the sister who creates the links between groups. The converse of this system can easily be imagined and does, in fact, exist among the so-called 'matrilineal Belu' of Timor. Among these Tetun-speakers, descent groups are structured on the categorical basis of relations between elder and younger female siblings while relations between opposite siblings are reversed. Brothers marry out (see G. Francillon 1967: 332ff). This case is all the more interesting because it is to the 'Belu' that the Rotinese seem to show the greatest linguistic and cultural affinities.

7. Marriage is not, however, the only means of creating alliances. Various sorts of long-term alliances of a political or ritual nature may also occur among clans or clan segments. Invariably these formal arrangements are attributed to a specific ancestral oath combined with animal sacrifice and possibly the drinking of blood. These may or may not have an effect on the regulation of marriage.

8. Life-cycle rituals follow a progression that goes from tender plant shoots to large hard-wood trees. Without a knowledge of their context, chants excerpted from these rituals might easily be mistaken for agricultural rituals (see Fox 1971b).

9. In this context, one may derive the simple symbólic equivalence: male: female::bone:blood::named spirit: physical person.

10. Increasingly, cash payments have come to be given in place of certain male objects. This may eventually transform the conception of this exchange as a set of obligatory prestations. Until now, however, cash payments merely stand for specific objects; i.e., 7,000 rupiah as 'sword'; 3,000 rupiah as 'spear', etc.

11. A certain amount of word play is involved in this dogma. In this context, the term for 'language' is also that for 'court-case'. The 'affairs of state' (*dedeä nusak*) are synonymous with the 'language of state'. This dogma, however, admits of one exception. The tiny domain of Bokai has suffered a sadly chequered history and is supposed to have been repopulated from outside. Hence poor Bokai has no 'language' of its own. See Fox (1974).

12. To make comparisons and to note general similarities among the domains is, in a sense, politically subversive. Each domain bases its claim to sovereignty and autonomy on a configuration of traditonal practices that are due to particular events which did not occur in other domains. For this reason, for example, each domain must maintain its own separate dynastic genealogy.

13. It should be noted that the royal lines in the oral genealogies of certain larger domains can be documented as correct and in proper sequence going back as far as 1660/1680. See Fox (1971a) for partial documentation of one such dynastic genealogy.

14. The population of Termanu and Thie:

	Area in sq. km.	1831*	1863	1921	1961
Termanu:	177	1,900	7,523	4,586	5,759
Thie:	93	3,400	8,136	7,911	12,116

*I am sceptical about the 1831 census as a valuable base line especially for Termanu. These figures were used to determine levies of men. They were taken before borders were clearly demarcated and, in the case of Termanu, just after a segment of the domain's population had been moved to Timor. I would suspect that Termanu's population is underenumerated.

15. Several of Termanu's original — 1831 — clans have diminished to the point that they are now no larger than a 'lineage' in another clan.

16. The clan and lineage structure of Termanu is complex. The analysis of it is made the more difficult by the existence of historical material that suggests some of the ways in which it took shape. Such material thwarts simple generalisations. I intend to discuss this structure in detail in *Termanu: The Political History of an Indonesian State*. Suffice it to say that there are actually three clans in Termanu that possess lineages with claims to noble status. These lineages — though not their clans as a whole — marry among themselves to a significant degree.

17. I am using the term 'moiety' here in a completely neutral sense as a 'division into two'. In many societies, as for example among certain Australian Aborigine groups, a moiety structure is primary. In other words, it is basic in that it marks the initial division from which all other divisions follow. Thie's view of its moiety system is just the opposite. The moiety system is the *culmination* of a long historical process. According to the genealogies and narratives, clans arose and segmented in irregular fashion and only after the firm foundation of the state, did one ruler call together all clans and get them to conduct sacrifices and swear oaths binding them in a moiety system. By all accounts, this event did not occur until well after the arrival of the Dutch. Instead of a primal division, native theory insists on that of a concluding contract. Unfortunately neither this native theory nor its opposite provides adequate explanation of the dual structures that one encounters on Roti.

18. There is another exception. One clan in Sabarai and another in Taratu, though under the formal jurisdiction of these moieties, are considered to be outside their marriage rules. Members of these clans may marry with any of the other clans. I discuss Thie's moiety system in detail in a forthcoming paper, 'Obligation and Alliance: State Structure and Moiety Organization in Thie, Roti' in Fox (ed.), *The Flow of Life*.

19. I think it is reasonable to describe marriage in Thie as a proscriptive rather than a prescriptive system. The sanctions against marriage, either within the clan or within the larger clan grouping, include a heavy fine, the requirement

of a waterbuffalo sacrifice, and the loss of bridewealth by the girl's family. The most severe sanction is the social stigma that applies to the children of such unions.

20. Briefly summarised, this system is attributed to the ruler, Thobias Messak (1844–1853) who established the payment of 5 waterbuffalo or 75 silver *gulden* (1 waterbuffalo = 15 *gulden*). This was divided into three named categories — *Beli-Inak, Oe-Ai,* and *Susu-Oe* — according to the recipient of these prestations — the category, *Susu-Oe*, for example, indicating the portion of bridewealth given to the mother's brother. The waterbuffalo were allocated as follows:

Beli-Inak:	2 waterbuffalo
Oe-Ai:	2 waterbuffalo
Susu-Oe:	1 waterbuffalo

This system was maintained until the reign of Johannes Messak (1882–1907) who, under pressure from members of his own clan and other nobles, created a two-tiered schedule of bridewealth. This schedule doubled bridewealth payments for noblewomen but retained the same payment to the mother's brother.

	Noble	Commoner	
Beli-Inak:	4	2	waterbuffalo
Oe-Ai:	2	1	waterbuffalo
Susu-Oe:	1	1	waterbuffalo

Salmoen Messak (1909–1917) rescinded this schedule and reestablished Thobias Messak's original schedule which is the one that is said to be in force to this day. Although this schedule is a major factor in bridewealth exchange, it has never been strictly observed. There is a euphemism — *Tuti-Uak*: 'to secure the welfare' (of the marriage) — that refers to secret bridewealth payments that violate the official standard. It should be noted, however, that these payments cannot be made in conspicuous exchange objects, such as waterbuffalo. They must be paid in cash in order to remain secret.

21. The official bridewealth schedule in Termanu is as follows:

4 waterbuffalo for daughters of the royal lineage
3 waterbuffalo for daughters of noble status
2 waterbuffalo for daughters of clan lords
1 waterbuffalo for daughters of all other commoners

This is not, in fact, a schedule to be interpreted literally. It is a statement of a system of rank that has originated outside of the exchange system but is maintained and employed by it. Although it can be considered as a symbolic scale (where — as has happened — 4 is taken to mean 40 waterbuffalo), it mainly provides high ranking bride-givers with justification for any level of bridewealth that they are able to negotiate.

22. The reference here is to Gullick's admirable discussion (1958) of lineage systems in western Malaya which bear many similarities to those of the Rotinese. The process involved here is also similar to what is described in the literature on 'conical clans'. I avoid this term only because of a certain ethnographic awkwardness I find in applying it to Termanu where only three

of nine clans are properly 'conical'.

23. One reason why this is not possible in Thie is because the mother's brother receives a *fixed* portion of bridewealth as laid down by the court schedule.
24. I should make clear that both MB (*Toö-huk*) and MMB (*Baï-huk*) are recognised in Thie but only the MB is entitled to specific payment. This is consistent with Thie's moiety system which shifts the emphasis of exchange from extended lines of relations toward a direct and balanced reciprocity.
25. This rough form of dating should not be taken to preclude the possibility that these ingredients were already present in Termanu at an earlier period.
26. I would like to qualify this statement by noting that the domains of Bilba, Ringgou, and to a lesser extent Korbaffo may have done something of the same sort as Termanu.

References

Chrijs, J.A. van der 1872. Koepang omstreeks 1750. *Tijdschrift voor Indische Taal-, Land- en Volkenkunde* 18, 209–27.

Fox, James J. 1968. *The Rotinese: a study of the social organization of an eastern Indonesian people.* Unpublished D. Phil. Thesis, Oxford University.

—— 1971a. A Rotinese dynastic genealogy: structure and event. In *The translation of culture*, T. Beidelman (ed.). London: Tavistock.

—— 1971b. Sister's child as plant: metaphors in an idiom of consanguinity. In *Rethinking kinship and marriage*, R. Needham (ed.). London: Tavistock.

—— 1974. Our ancestors spoke in pairs: Rotinese views of language, dialect, and code. In *Explorations in the ethnography of speaking*, R. Bauman and J. Sherzer (eds.). Cambridge: Cambridge University Press.

—— 1977a. *Harvest of the palm: ecological change in eastern Indonesia.* Cambridge: Harvard University Press.

—— 1977b. The textile traditions of Roti, Ndao and Savu. In *Textile traditions of Indonesia*, M. Kahlenberg (ed.). Los Angeles: Los Angeles County Museum of Art.

—— (in press). Obligation and alliance: state structure and moiety organization in Thie, Roti. In *The flow of life: essays on eastern Indonesia,* J.J. Fox (ed.). Cambridge: Harvard University Press.

—— (in preparation). Termanu: the political history of an Indonesian state.

Francillon, G. 1967. *Some matriarchic aspects of the social structure of the southern Tetun of middle Timor.* Unpublished Ph.D. Thesis, The Australian National University, Canberra.

Gullick, J.M. 1958. *Indigenous political systems of western Malaya.* London: The Athlone Press.

Leupe, P.A. 1877. Besognes der hooge regeering te Batavia, gehouden over de commissie van Paravacini naar Timor 1756. *Bijdragen tot de Taal-, Land- en Volkenkunde* 25, 421–94.

Mauss, M. and H. Beuchat (in press). *Seasonal variations of the Eskimo: a study in social morphology.* (Translated with an introduction by J.J. Fox.) London: Routledge and Kegan Paul.

Ormeling, F.J. 1956. *The Timor problem.* Groningen: J.B. Wolters.

SAGO SUBSISTENCE AND THE TRADE IN SPICES:
A PROVISIONAL MODEL OF ECOLOGICAL SUCCESSION AND IMBALANCE IN MOLUCCAN HISTORY

ROY F. ELLEN

It is now abundantly clear that to characterise the subsistence pattern of 'outer' Indonesia as swidden cultivation closely linked to a prevailing tropical forest ecosystem (Geertz 1963: 12–37) is unsatisfactory. Swiddening is certainly much more variable than Geertz allows for in his model, in its technical practices, social context and environmental relations (Spencer 1966). Moreover, some communities rely to a large extent on the collection of non-domesticated food and materials (Ellen 1975), while certain ecologically distinctive palm economies have a surprisingly widespread distribution (Barrau 1959; Ellen 1975; Fox 1977; Johnson and Ruddle 1976). Among the most complex and historically important of these is the traditional mode of subsistence of the Moluccas[1] focussed on the appropriation of starch from the *Metroxylon* palm. This paper examines its relationship to growth in the production and trade in nutmeg, mace and cloves during the sixteenth and early seventeenth centuries.

I begin by outlining the general hypothetical structure of the economy of communities not involved in the production of spices or other commodities on a significant scale. This initial construct is designed to apply to all Moluccan economies prior to their integration into higher level systems involving regional and international trade. It therefore applies equally to pre-fifteenth century north Moluccas, pre-sixteenth century central Moluccas and those contemporary isolated subsistence economies not yet fully integrated within wider systems. On the basis of this, the model is elaborated to account for certain distinctive features of the process of successive inclusion within wider trade systems. This is done by modifying a number of general characteristic processes of socio-economic change in the light of what is known of Moluccan society since 1500.

The Model

I want to suggest that the properties of *Metroxylon* sago as a subsistence resource are intimately linked to fluctuation and expansion
in the trade in spices in the Moluccas over the course of several
hundred years. The model assumes that the elementary Moluccan
subsistence unit (hereafter referred to as EMSU) was progressively
drawn into increasingly wider exchange systems and, as a result,
larger and more complex modes of production and systems of total
production (see Friedman 1976). For convenience the process is
divided into four phases: the early, formative, mature and late
exchange phases.

The first of these is dominated, for the most part, by simple
relations of production associated with the EMSU. During this phase
small amounts of spices are collected and exchanged for a variety of
material items, in particular valuables and usually excluding food.
This process steadily accelerates towards the end of the phase as
external demand for spices and internal requirements for the valuables
of traditional local exchanges increase.

The formative exchange phase is recognised by an upswing in the
level of production and the full domestication of spice trees. Trade
becomes of increasing political significance to both producers and
traders. The control of the growing volume of commodities by
individuals and small kinship groups forms a growth-point for class
differentiation. The process is amplified through external support for
indigenous rulers through whose hands the trade passes.

The mature exchange phase is marked by a further increase in the
scale of cultivation and, most importantly, growth in the local trade
of foodstuffs, particularly sago, along traditional communication
routes into areas hitherto self-sufficient. In this way further EMSUs
are drawn into the trade system. There is a corresponding growth in
the population of spice-producing areas. Growth continues in spice
production, local trade in food staples and population. The pattern
is consolidated and the communities are drawn within a global
exchange sphere of which they are a quasi-dependent element.

This phase of consolidation is characterised by two self-amplifying
loops of dependence, more generally a feature of those emerging
relations of unequal exchange which result in underdevelopment.
These are illustrated in Figure 1. In the first (a) the increasing area
devoted to production for exchange leads to more intensive use of
available land for subsistence crops but eventually a net decrease in
land available and output of production for use. This in turn leads to
more reliance on production for exchange and consequently to a

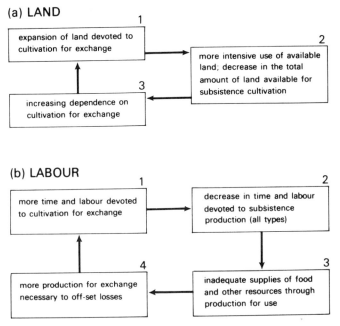

Fig. 1. Self-amplifying loops in the process of integrating local economies into wider systems of exchange.

further expansion of production activities in this sector. The second loop (b) is functionally similar but concerns human effort. As more time and labour are devoted to cultivation for exchange, so there is a drop in subsistence production. Under other conditions this might be directly related to increasing inputs of labour in the exchange sector. However, the increased effort required in this case is minimal. What seems most significant in accelerating the cycle is the relatively higher return for labour compared with subsistence production. This may be partly modified by shifts in technical arrangements and in the social division of labour, but ultimately decisions are made which result in the allocation of progressively less effort to subsistence crops. This results in a drop in the level of production for use, which in turn leads to more production for exchange to make up losses of food in the subsistence sector.

Now while these loops may be operating in their classic form in particular localities and at the level of gross generalisation, the traditional Moluccan pattern of subsistence based on sago and the peculiar trade nexus which developed significantly altered their shape

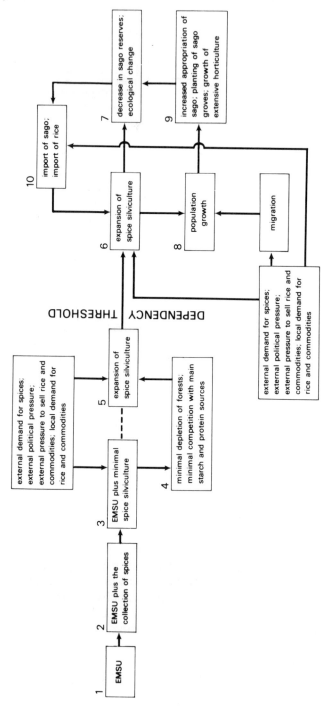

and the rate at which they were circuited. Subsistence on sago delayed and altered the character of the dependent relationship by preventing the immediate depletion of arable land, rapid deforestation and competition over land between subsistence and trade crops. The relative drainage of resources in particular areas was offset by local trade in sago along traditional pathways. It was only the intervention of European mercantilism and capitalism to break this trading network which increased the speed of change in the direction of depen dependency and resulted in chronic ecological instability. These processes are summarily diagrammed in Figure 2, loop 3-4-5 representing the potential ability of the system to delay the onset of dependency.

The late exchange phase is characterised by a rapid decline in the trade in spices in the context of an artificial control of spice production and trade. Local trading is resumed, alleviating to some degree the problems of unequal resource distribution. The system becomes a relatively stable but deteriorating combination of dependency and self-sufficiency until the mid-nineteenth century. From this period onwards, there is a gradual resumption and geographical extension of the processes dominant in the formative and mature exchange phases.

The Elementary Moluccan Subsistence Unit

The main source of carbohydrate in the Moluccas is the starchy flour of the palm *Metroxylon sagu*, the sago of commerce. Sago is also known and extracted from *Arenga pinnata, Areca catechu, Nypa fruticans, Licuala rumphii*, and also other species. But only in a few areas do they constitute more than famine and occasional foods owing to their fibrous marrows, unpalatable pith and arduous processing (van Slooten 1959: 335). One such area is Halmahera (Fortgens 1909). The natural sago resources of the area are considerable, comprising several thousands of square kilometres of otherwise useless swamp forest, mainly distributed on the islands of Seram, Halmahera, Buru, Aru and Bacan (Ruinen 1920). For the purposes of this discussion, it is also necessary to include the even more extensive sago resources of Teluk Berau, Misool and Salawati in the Vogelkop area of Irian Jaya (Indonesian western New Guinea). (See Figure 3.)

Among the Nuaulu of south central Seram[2] sago provides some 1,958 Cals per adult per day, 63% of the entire energy intake. This

---

Fig. 2. (opposite) A general model of aspects of ecological and economic change in the Moluccas.

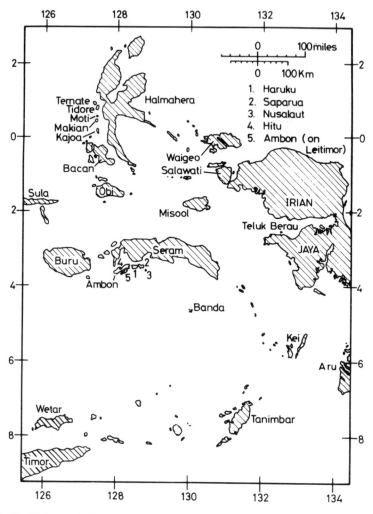

Fig. 3. The Moluccas, indicating places mentioned in the text.

is somewhat lower than recent figures obtained for two Papua New Guinea societies where sago is of central dietary and economic importance. Ohtsuka (1976: 4) has recorded a figure of 70% for a people of the Oriomo plateau, while Townsend (1974: 230) has recorded 85% for the Sanio-Hiowe of the Upper Sepik region. Comparative observations in the Moluccas and statements in the literature suggest that such figures are very likely to be reached in parts of the area under consideration, and that generally sago constitutes around

60% of all carbohydrate consumed. Although there are certainly local exceptions to this pattern, it is clear that it has been the primary one in the region at least since the fifteenth century and remains, technically, virtually unchanged (Galvão 1544: 43, 45, 133–37; Dames 1918–21: 202 following De Barros and Pigafetta).

Metroxylon sago is extremely productive in terms of yield per palm. In south Seram 100 cm³ of trunk provides on average about 0.02 kgs of moist flour, which means that a good palm can yield up to 300 kgs (the equivalent of 106 x 10⁴ Cals) (Ellen 1975: 142). This high calorific equivalent is due to the fact that the amount of raw starch may rise to 84–85% of the flour itself (Burkill 1935: 1462).

In general, estimates of sago yields per palm from *Metroxylon* in Southeast Asia vary from 25 to over 360 kgs, and there is considerable local variation (Johnson and Ruddle 1976: 11). Burkill (1935: 1461) reports yields of between 113.4 and 299.4 kgs and mentions that on occasions yields may reach 544.3 kgs. More recent and independent sources indicate very high yields (over 200 kgs) from places as far apart as Mindanao, the Sepik, West Malaysia and Sabah (Johnson and Ruddle 1976: 11). It is interesting that over a century ago Wallace had recorded yields for Seram, in particular, of 272 kgs (1869: 192).

Estimates of *Metroxylon* productivity per hectare vary enormously, from 7 to 330 palms, depending on maturation period, time of harvesting and whether the palms are cultivated or wild. But given that sago usually grows in swampland that is otherwise lost to production, sago appropriation makes very economical use of land resources.

Estimates of productivity in terms of output per hour are much more likely to be useful and reliable. 2.2 kgs per hour has been quoted as an average figure for the sago areas of the Sepik (Johnson and Ruddle 1976: 10). On east Seram, Wallace recorded output of about 40 kgs per day (1869: 292). However this figure is interpreted, it remains high. Output per man-hour among the Nuaulu is around 2.6 kgs (Ellen: fieldnotes). If variation is due to technical apparatus, division of labour, work organisation and the starch content of different palms, then these may combine in the Seramese case to explain higher productivity per man hour. Certainly, there is evidence that the productivity of palms is higher, that the techniques are more efficient than those used in the Sepik and that the labour input is generally male rather than female. On the whole, it is clear that sago provides more calories per man hour of labour, sometimes as much as three times, than domesticated tubers and grains (Ohtsuka 1976: 10).

Clearly there are certain disadvantages in relying on sago to such an extent and specific difficulties related to its extraction (*ibid*.: 10–11), but it remains an extremely useful food source in the context of Moluccan ecology, economy and population distribution. Apart from the qualities mentioned, it has year-round availability (many tubers and grains are seasonal), and it suffers little from damage caused by wild animals, other pests and abnormal climatic conditions (*ibid*.). Moreover, *Metroxylon* palms have a wide range of additional uses as a source of liquor, tinder, utensils, containers, construction materials, medicines and such like (Burkill 1935: 1462–3; Ellen 1976). The Moluccan sago economy is no less remarkable in this sense (and probably more so) than the Rotinese and Savunese lontar palm economy (Fox 1977).

One important consequence of extensive reliance on sago in Moluccan economies is that the normal cultivated sources of carbohydrate – tubers and grains – while grown in small quantities, are of less importance. As a result, the amount of land required for cultivation of subsistence crops is less than in societies which are almost totally dependent, say, on rice or yams. It seems that sago subsistence has had the effect of limiting horticultural development (Barrau 1958: 14), maintaining it at a low level or making it altogether unnecessary. There is little incentive to alter a landscape at some effort and adopt new and untried techniques when a living can be had by simply extracting resources from an unmodified environment (cf. Forrest 1779: 42).

It is highly likely that many Moluccan communities were, in the relatively recent past, either not dependent on cultivated plants at all or only minimally so. Even today, the contribution from all non-domesticated resources is significant, over and above the very special position of sago (Ellen 1975). Among the Nuaulu, the dietary contribution of non-domesticated resources is substantial. It provides around 1,272 Cals of energy per head per day, compared with 1,823 Cals derived from all domesticated sources; that is, over 40% of an estimated daily adult intake of about 3,085 Cals. Non-domesticated sources also provide 64% of the protein fraction of all consumed food, as well as being of paramount importance in terms of fat-content.

This food value is obtained from around 120 species of animal and 48 species of plant. But although the total number of non-domesticated species used in this fashion is extensive, over 80% by weight of plant-derived food comes from *Metroxylon*. The high carbohydrate input is almost solely due to the importance of naturally-reproducing sago in Nuaulu economy. Similarly, almost 90% by weight of animal derived foods come from pig, fish, cuscus and deer. Consequently, despite apparent variety, the principal sources of food energy and protein are extremely specific.

Nuaulu male labour involved in the appropriation of non-domesticated resources exceeds that involved in working domesticated resources, being at lest 56% of the total energy expenditure devoted to productive activities (*ibid.*: 144). Female labour involved in the appropriation of non-domesticated resources is less, though not by much. In the central highlands of Seram, horticultural activity appears to be even less than among the Nuaulu (Valeri, pers. comm.), and this situation probably extends throughout much of the interior. On Halmahera, Buru and the Aru islands the situation is similar, or at least was so until recently (Campen 1885: 8–12; Riedel 1885; Bosscher 1853). Such food collecting and sago economies were almost certainly much more widespread in the past.

Forest resources have historically also been important in exchange in these same areas and were probably the first items to attract outside merchants: rattan, damar resin, wax, timber, feathers and other animal products. It is exchange of this kind that must have introduced to the Moluccans the idea of regular trade in cloves and nutmegs, for these also were originally gathered from wild trees. The extent of modern and historical trade in forest products, in which the rural communities themselves were intermediaries, is clear from colonial accounts, regional economic histories and trade figures (see Dunn 1975; Cobban 1968).

However, it is likely that since approximately 1400 the extent of plant domestication in the Moluccas and the form this has taken has changed radically. For one thing, sago itself has been increasingly domesticated. In many areas, particularly the more densely populated parts of the smaller islands and along the coasts, much of the sago comes from tended groves. 46% of all calories obtained from sago by the Nuaulu come from domesticated palms, and 48% of the energy expended in activities connected with sago appropriation concern such palms (Ellen 1978). However, the distinction between domesticated and wild palms is often difficult to make since the former are often left to vegetatively propagate themselves, while otherwise wild palms are protected and new palms in wild areas planted from basal suckers. Planting in general is haphazard and wild and transplanted stands may coexist in groves. A similar practice is found in relation to other trees, such as species of *Pandanus* and *Canarium*, and including *Eugenia aromatica* (the clove) and *Myristica fragrans* (the nutmeg).

What cultivation did exist in the Moluccas prior to the fifteenth century was probably mainly connected with species of yam, banana and taro. From the sixteenth century onwards, a wide range of cultivars which had hitherto been either absent or peripheral became more important. Among these were manioc, *Xanthosoma*, maize, sweet potato and rice, and their increasing importance can be directly related to the growth of political and economic contact on a regional, and then international, scale. This obviously affected the degree to which many Moluccan communities were involved in horticulture, but sago was evidently not replaced as their primary food source.

The heavy reliance on non-domesticated resources is inevitably reflected in carrying-capacity calculations. For example, very little cultivated land is required to provide for a hypothetical average Nuaulu individual: just 0.25 hectares. The population supportable in the same locality has an upper limit of $200/km^2$. This is exceptionally high compared with the actual densities of other groups of swidden horitculturalists. Similarly, the smallest area of cultivable land capable of supporting the village of Ruhuwa (180 persons) was minimally 74.9 hectares. In Southeast Asian terms, the total productive cultivated land, high-energy yield cultivated land and the total potentially cultivable land are all low (Ellen 1978).

Summarising, our model EMSU can therefore be characterised by
a) a critical dependence on sago as a source of carbohydrate; and
b) a minimal use of domesticated resources.
To this we must also add
c) minimal reliance on outside communities for food and materials — minimal production for exchange;
d) relatively simple techniques and technical relations of production; and
e) social relations of production realised through ties of kinship, affinity and locality.
For our purposes, then, the elementary Moluccan subsistence unit plus its associated social relations constitutes a distinct mode of production and total system of reproduction, in that it has a high degree of systemness and autonomy. In this, and also in terms of its widespread distribution and archaic character, it contrasts with, say, Rotinese and Savunese lontar palm economy. Although it may indeed have ancient origins, the latter is largely an historically recent adaptation to environmental depletion (Fox 1976).

The Production of Sago for Exchange

Sago is, and always has been, predominantly a subsistence crop. However, it is likely that from the earliest times of its production it has been traded in the Moluccan region. For despite what has already been said, the distribution of edible sago is uneven and does not neatly fit the pattern of human settlement. The smaller volcanic islands are less suited to sago swamp forest; inhabited atolls and sand banks can support little vegetable matter at all, and areas that are environmentally suitable may still have insufficient resources to meet requirements. Such localities had to import sago, and in certain cases, other foodstuffs as well. They were able to do this by exchanging locality-specific produce (fish, shell, stone, forest products). In some places, craft specialisations probably emerged — pottery on Saparua, Haruku, Ambon, Kei and Marē (Ellen and Glover 1974), the production of shell bracelets on Goram, cloth in west Seram and polished stone axes elsewhere. This kind of local trade involving sago is well documented for Melanesia (e.g. Barrau 1958: 15; Brookfield 1971: 314–34; Sahlins 1972: 277–314; Allen 1978) and, in the Moluccas, must certainly antedate organised trade with outsiders. It is conceivable that such exchanges formed the basis for the emergence of centralised and stratified polities on the smaller islands, which were in a position, because of the balance of trade, to extend their authority in some form over wide areas on other and larger islands.

There are a number of characteristics of *Metroxylon* sago which make it eminently suitable as an item of trade. We have already noted that it is a particularly concentrated form of starch. As a flour, it is easily transportable. Damp, it ferments slightly and this allows it to be stored for periods of up to several months. Baked into hard biscuits, it has phenomenal storage properties and may retain its edibility for many years (Galvão 1544). Its storage potential and compactness has also made it an important item in New Guinea trade networks (cf. Townsend 1974: 133).

So sago has been traded locally on a regular basis for over five hundred years, to offset local short-term deficits and to supply areas of chronic shortage and total absence. However, it seems that the trade in sago reached higher levels from the fifteenth century onwards.

The Moluccan Trade in Cloves: 1450–1700

There is evidence from Europe and China for the trade in Moluccan spices stemming back more than 2,000 years, but the early history is fragmentary. By the twelfth century, the economic importance of the Moluccas was substantial, being the source of nutmeg, mace and cloves and in return a market for rice, textiles and manufactured goods from India and elsewhere. By the beginning of the thirteenth century, Moluccan spices were being handled in Srivijaya, probably arriving via the towns of north Java. In the fourteenth century, the trade in spices appears to have grown sharply, largely due to European demand. By this time, the trade was virtually dominated by merchants from the Javanese seaports; these ports had by then been incorporated within the Hindu kingdom of Majapahit, which is reputed to have extended its authority as far as the Moluccas. This may have been partly to ensure the supply of spices from Banda and the north Moluccas and may have been accompanied by the levying of tribute and effected through the exchange of valuables and food-stuffs (Meilink-Roelofsz 1962: 14, 23).

Cloves first appear to have been actively collected for exchange on the coastal lowlands of Halmahera and the small islands immediately to the west — Ternate, Tidore, Moti, Makian and Bacan. It is likely that this was also the centre of its speciation, although indigenous claims that it first originated on Makian (Van Slooten 1959: 321, following Rumphius) are untestable. In fact, clove trees appear to have been widespread in the forests of the northern and central Moluccas, although confined until quite late to this area. There may be environmental reasons for this, since the clove tree seems to require certain specific conditions to flourish and these are met in the Moluccas. It grows best under a rainfall regime of 127–178 cms, on deep moist rich loamy soil at altitudes up to 600 metres and 'within sight of the sea' (Cobley 1956: 204). The rich sandy volcanic loams of the northern Moluccas seem particularly suitable, although the tree also flourishes on loams of disintegrating sandstone and secondary rock on Ambon.

At the beginning of the sixteenth century clove production was concentrated on the islands of Ternate, Tidore, Moti, Makian and Bacan (Barbosa 1518: 109), and we have some surprisingly full accounts of the technical and social arrangements involved (e.g. Galvao 1544: 137–39). Technically, production has remained virtually unchanged down to the present time. The forests appear to have been full of clove trees, and cloves were still largely gathered from these wild trees rather than cultivated in groves. On Bacan, cultivation was recent, and on Tidore more recent than Ternate, although there is no suggestion in the sources as to how archaic clove culture was there (Meilink-Roelofsz 1962: 98, 155). Makian,

with its good harbour, was an export centre, also importing cloves
from other islands. Tidore had no harbour suitable for long-distance
vessels. Only Ternate exported its own cloves, which may in part
explain its importance. Cloves gathered from the large number of wild
trees on Halmahera were exported via the port of Gilolo, and there is
a suggestion in Pires that cloves were just beginning to be cultivated
there as well (1512–15: 213–221).

The social context in which all this took place was of emerging
centralised and stratified polities. Prior to the rise of the clove trade,
the societies of the northern Moluccas must have conformed pretty
much to the EMSU pattern, and what fragmentary evidence we do
possess bears this contention out (Galvão 1544: 77–79, 105, 133).
The small islands off the west coast of Halmahera were probably of
relatively little significance, although they must have been partially
dependent on the import of food and materials and this may have
encouraged early economic specialisation and required an element of
political control over trade routes. Such locations are also those
characteristically chosen by intrusive merchants, since they have easy
access and yet are relatively well protected (Ellen and Glover 1974:
368). The first merchants must have been relatively mobile, and
markets periodic and irregular.

As the scale of the trade increased, so permanent central places
must have arisen. However it is difficult to understand the emergence
of places like Ternate, Tidore — and later Banda and the centres of
the central Moluccas — in terms of classic central place theory; the
usual assumptions do not hold. In fact, disturbances created by
various nonisotropic conditions — variations in population density,
demand, geographical irregularities, endemic warfare and differential
distribution of transport facilities — have tended to shift the assumed
location of centres (cf. Smith 1976: 28–30).

At the time of Portuguese contact, the lucrative trade with outsiders was very much in the hands of the northern Moluccan rulers
(Meilink-Roelofsz 1962: 97), and it is possible that their participation
in commerce as intermediaries helped consolidate their sovereignty
(Galvão 1544: 77–9, 105, 133). A considerable part of the clove
crop went as tribute to rulers, who could then provide foreign
merchants with larger quantities (Meilink-Roelofsz 1962: 157). By
this time also, the role of rulers as intermediaries in the trade with
Malays and Javanese Muslims had probably encouraged a degree of
Islamisation of the elite, although the process was by no means
complete (Barbosa 1518: 203; Pires 1512–15: 213). The Muslim
community must have also extended to other persons with direct

contact with traders and foreign merchants living in coastal settlements.

The process of political evolution appears to have reached its peak with Ternate and Tidore, whose rulers were hostile to each other and who both claimed other parts of the Moluccas. Both maintained close links with centres on Halmahera. Moti appears to have been equally divided between them. Ternate especially extended its area of influence during the sixteenth century, and at its zenith this included Bacan, Mindanao, parts of Sulawesi, Solor, Banda, parts of Seram, Ambon and elsewhere. It was this that in part gave rise to conflict with the Portuguese (*ibid.*: 215–18; Meilink-Roelofsz 1962: 154). By 1534, Tidore had conquered Misool and exercised control over parts of east Seram. Nearer home, it appears to have had some influence over Makian. Both Ternate and Tidore had influence over parts of New Guinea (Elmberg 1968: 125).

At the beginning of the sixteenth century, most of the trade appears to have been with Javanese and Malays, and to a much lesser extent Chinese. The Malaccans were undertaking regular trading voyages between 1511 and 1522, calling sometimes at Ambon and sometimes in the northern Moluccas or Banda for the purpose of taking on cargo. The Bandanese seem to have done a good deal of intermediary trading, reselling Moluccan cloves to Javanese and Malays in the Banda islands. Other than the Bandanese, the Moluccans seem to have had no merchant shipping of their own. Although their oar-propelled *kora-koras* were suitable for local trade and warfare, for the transportation of bulk produce to the outside world they were dependent on foreign merchants (Meilink-Roelofsz 1962: 88, 97, 99–100, 153, 158, 160).

The clove trade was based entirely on barter and the terms were clearly exploitative of the local inhabitants. In exchange for cloves, the northern Moluccans received copper, quicksilver, vermillion, Cambaya cloth, cummin, some silver, porcelain and Javanese metal gongs from Malays and Javanese. From the Chinese, they received Javanese cloth, silver, iron, ivory, beads and blue earthenware and porcelain dishes and cups (*ibid.*: 99, 158). Chinese coins were also apparently imported via Malacca (Barbosa 1518: 175, n. 1; Dames 1918: 202, n.1, following De Barros; Galvão 1544: 79, 139). Goods also reached Ternate from elsewhere; for example, iron axes, swords and knives came from the Banggai archipelago, and from other places some gold, ivory, coarse native cloth and parrot feathers (Pires 1512–15: 215–16). Porcelain, cloth and metalwork seem to have already been established as important items in local trade and ceremonial exchange by this time.

Later the same pattern was to be repeated in the central Moluccas, with the main items exchanged for cloves being Cambaya and other cloths, which were an important part of local prestige patterns and ceremonial exchange, and also foodstuffs (Barbosa 1518: 99; Galvão 1544: 141).

The objects found in modern settlements as valuables, deposited in museum collections or recovered archaeologically, indicate the antiquity of the trade, and also the importance of the objects in traditional ceremonial in even the most isolated areas, including New Guinea.

The volume of trade in cloves was enormous, and continued to grow throughout the sixteenth century. Actual production figures for the period are, however, difficult to interpret, Those of Pires, Rebello, Carrion, De Brito and Corneliouszoon are generally agreed to be gross over-estimates, even if it is assumed that they refer to the big

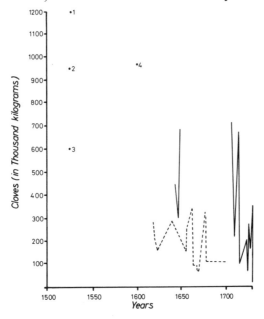

Fig. 4. Some approximate measures of output and trade in Moluccan cloves between 1500 and 1729, measured in thousands of kilograms. The numbered estimates of total production are those of (1) Urdaneta (1526–35), (2) Ovieda, (3) Pigafetta and (4) a mean of good and bad years from a Portuguese statement (Van Leur 1967: 368 n.18; Meilink-Roelofsz 1962: 352 n.81). The continuous line represents cloves acquired by the Dutch between 1645 and 1648 (Bassett 1958: 15) and Moluccan harvests (1709–1729). The broken line represents a combination of Dutch sales and offers on the European market (1619–1700). The last two series are based on Koloniaal Archief sources used by Glamann (1958: 93–110).

harvests which occurred every three to seven years (Meilink-Roelofsz 1962: 97, 352 n.81; Galvão 1544: 137). Calculations are also hampered by the difficulties in reducing variable Asiatic measures to European equivalents.[3] However, in Figure 4 I have used four of the more reliable figures available for the sixteenth century and placed them alongside later Dutch East India Company figures for European sales, auction offers and actual production. By contrast, these figures are likely to be underestimates of total production, and even sales, since the Dutch were also selling in Asian markets while a substantial fraction of the crop continued for some time to be handled by non-Dutch traders. However, the sixteenth century trade was considerable by later standards. Bickmore (1854: 155) calculated, using Pigaffetta's figures, that a conservative estimate of the 1521 output was 17 times that for the middle of the nineteenth century. But if the actual trade was not as voluminous as the more extravagent estimates suggest, then production certainly exceeded actual sales. Barbosa (1518: 202) describes the quantities produced in the northern Moluccas as being so large that they could not all be exported; many remained unpicked on trees. This suggests that local demand for imported valuables was high and growing, probably assuming more importance in ceremonial exchanges, but it may also be related to the demands of rulers to meet their own exchange obligations and consolidate their positions in the process of class formation. Such overproduction, of course, also worked to the advantage of traders in keeping the cost of cloves low.

Production and trade in the northern Moluccas continued to grow throughout the sixteenth century, although there were local variations. By 1599, for example, Bacan was no longer of any consequence, and by 1606 the most productive of the northern islands was Makian (Meilink-Roelofsz 1962: 184, n.81).

However, by the 1620's clove production in the northern Moluccas had dropped dramatically. In 1623, the islands were supplying practically no more cloves at all (*ibid.*: 221). This has been attributed to the desire of Ternate and Tidore to rid themselves of European merchants. It may also have been affected by the age of the trees and by the fact that they were no longer being planted because of a great decline in the population through heavy war fatalities and periodic epidemics. Bacan appears to have been particularly badly affected (*ibid.*: 216). The centre of gravity of production and trade had shifted to the central Moluccas, while Ternate and the north began to rely almost entirely on local trade, particularly with New Guinea.

During the early part of the sixteenth century, the central

Moluccas appear not to have played a significant role in the trade or cultivation of cloves (Pires 1512–15: 212), although the Ambonese seem to have been involved in handling some trade between Java and the Moluccas (Tiele 1879: *3*, 23, after Castanheda). Ambon itself appears to have been under the influence of the northern Moluccan states (Dames 1918–21: 200, n., after Castanheda and De Barros). It was used as a watering place and provisioning point for boats bound for the Moluccas (*ibid.*: 199, n.1, quoting Linschoten). Perhaps we must assume that local exchange and prestige were not so dependent on non-locally produced valuables. But as demand for spices increased, so production spread southwards. There is evidence that at this time large parts of the central Moluccas were at last moving into the early exchange phase, and some cloves (probably still uncultivated ones) went from here to Banda (Meilink-Roelofsz 1962: 100). Clove cultivation cannot have been introduced to the central Moluccas before the beginning of the century, since there is no mention of clove among the crops listed by Pires and since, at the time of the Dutch arrival, the clove trees at Kambello on Seram were the first to have been planted. It seems that cloves had first been introduced from Makian to Seram and then to Ambon (*ibid.*: 159).

The growth in the importance of the central Moluccas was rapid due to increased Portuguese and Asian demand. Javanese influences, particularly, were visible by the middle of the century, by which time Ambon was rich in cloves (*ibid.*: 160–161, 25). Tiele (1879: 435, following Conto) says that by 1566–70 the Javanese were taking 1,000 quintals of cloves from Ambon but that there was a total crop of 2,000 quintals (117,760 kgs). Clearly, overproduction stimulated by local demand, and a poor understanding of the functioning of the market, was present in the central as well as the northern Moluccas (Meilink-Roelofsz 1962: 100).

The gathering and then the cultivation of cloves spread to an increasing number of islands in the course of the sixteenth century. In places where clove did not occur wild, it must have been intro-duced as a domesticate in the first place. The central Moluccas moved rapidly through the formative period and, by the early seventeenth century, many settlements were relying heavily on imported food-stuffs, were producing little or no sago and, with extensive groves of clove trees, were experiencing growing population densities and rapid deforestation.

The centre of the clove trade on Ambon became the Muslim village of Hitu, over which Ternate also claimed to rule, together with the rest of Ambon and Seram. The Ambonese and Seramese were not involved in the bulk transport of cloves themselves, although with Javanese helmsmen they were involved in fetching slaves, iron knives and swords from Banggai, Buton, Salayer and elsewhere. Although the original settlement of Hitu cannot have been connected with the spice trade alone, since it predates the period of clove production on either Ambon or Seram, its development as a political and market centre must certainly be associated with it. However, Hitu did not grow cloves itself but received them as tribute from smaller places, which stood in a dependent relationship and were obliged to provide certain commodities other than cloves. Conflict between the Portuguese and Hitu and its Javanese and Bandanese allies led to the promotion by the former of clove cultivation on Leitimor (*ibid.*: 100, 160–61, 220).

By the first decade of the seventeenth century, there were a number of important clove villages on Seram, for example Luhu and Kambello (*ibid.*: 201). Gradually, gathering, cultivation and trade extended eastwards to include such centres as Hatusila on Elpaputih Bay and Lessidi (Tiele 1878: 439) although clove cultivation was restricted environmentally to predominantly coastal areas below an altitude of 600 metres. As demand rose, there was initially a greater incentive to protect and tend existing wild trees and seedlings, then to plant from seed, and finally to grow them from seed in special groves. This process has been repeated at the present time for more isolated communities as they become involved in cash cropping. In the right locality, the process was not complicated by the necessity of acquiring complex new techniques or social organisation. Wild trees could be protected; young trees could be grown from seed in nurseries or at the end of water conduits, and transplanted into depleted forest, in swiddens or at their edges, or in areas cleared for the purpose. In exposed dry areas, however, many seedlings were likely to have died in the process (cf. Ellen 1978).

By 1630, expansion in the central Moluccas had been considerable, largely stimulated by the intervention of English, Danish and particularly Dutch traders. Trees were planted even more intensively and over a widely scattered area. Competition led to increased output, and in 1622 the quantity of cloves grown on Ambon and Seram was double the world consumption of this product (Meilink-Roelofsz 1962: 218, 224).

The Local Trade in Food Stuffs

As we have seen, the centres of the spice trade were for the most part on small islands. As clove production and population expanded, they became progressively incapable of satisfying their own food requirements.

There are suggestions in the literature (e.g. *ibid.*: 23, 138; Masselman 1963: 460–1; Van Leur 1955: 128–9) that losses in subsistence production were made up through the import of rice from Java. Now it is clear that the quantities of rice being transported from Java to the Moluccas were large at this period, although Van Leur's attempts to calculate the volume of trade from the assumed consumption of Ambon, Banda and the northern Moluccas must surely be spurious. Without doubting the scale of trade, or its importance to the Javanese towns, it is unlikely ever to have been sufficient to meet more than a fraction of Moluccan subsistence requirements. Today, with a more urbanised, immigrant and acculturated population, total official rice imports to the Moluccas are only 16,534,626 kgs. Divide this by a population of 1,080,874 and the average amount of rice per person is 15 kgs a year (Kantor Sensus dan Statistik 1972, 1974). If all Nuaulu calorie requirements were to be met by rice, they would need 236.66 kgs per year per person (calculated from data presented in Ellen 1978). Of course, the import figure is for the entire province, and localities such as Ambon, Banda and Ternate imported, and continue to import, much larger quantities per head of population. In 1911, for example, the entire northern Moluccas imported 32.14 kgs per head (ENI 1917–35: *4*, 314–22). Current rice production in the Moluccas is negligible. Although it is recorded that rice was grown on Moti, Halmahera and elsewhere in the sixteenth century (Pires 1512–15: 217; Galvão 1544: 133), and may have been grown in small quantities even much earlier (cf. Glover 1977), it is unlikely to have been cultivated on any large scale for environmental, technical and social reasons. It is clear that then as now, the bulk of the food had to be derived through subsistence appropriation or local trade, and given what is known of Moluccan patterns of subsistence and ecology, this must have consisted largely of sago. Imported rice was probably mainly destined for the small immigrant communities and the burgeoning ruling elites, associated with an imported elitist culture and more widely used as a prestige (rather than ritual) food for special occasions (de Clercq 1890). This pattern of consumption has been more-or-less maintained to this day (Ellen: fieldnotes).

I have indicated that most of the early sources suggest that sago was one of the most important products of the Moluccas at the time of the arrival of the Portuguese, and references to 'bread' in the same literature almost certainly refer to sago biscuits (Galvão 1544: 133–37).

Although swiddening and dry field horticulture were present (*ibid.*), because native Moluccans were not dependent upon it, clove groves could expand without competing for space with other crops and be planted in the localities most suited to them. Also, the main subsistence crop (sago) and main exchange crop (cloves) did not compete for land; they occupied completely different niches in the local ecology. Moreover, sago and, even more so, cloves did not require large amounts of labour, compared with horticulture or irrigated agriculture. This is a general feature of silviculture which also delayed dependency until output had reached a much higher level than is otherwise necessary. Cloves were accepted, and expanded rapidly, precisely because they did not initially threaten subsistence, and brought increased trading opportunities to the inhabitants at little cost. Expansion was encouraged by the rulers of petty states, enabling them to consolidate their position in the emerging class structure and exercise control over the surrounding areas.

But the smaller and, increasingly, more densely populated clove-producing islands could not produce sufficient food to meet their needs. Sago had to be imported from elsewhere (Van Leur 1967: 369 n.18). Similarly, growing population centres – such as Hitu – had to import food from the surrounding countryside. In the northern Moluccas, Ternate and Tidore, although producing a little food (including sago), were net importers. A measure of their dependence were the shortages incurred during times of war (Galvão 1544: 231). At the beginning of the sixteenth century, the possession of half the island of Moti apparently ensured the ruler of Ternate an adequate supply of food for the needs of his own subjects (Pires 1512–15: 214–15). According to Pires, food crops were grown on Tidore, possibly including sago (*ibid.*: 217), but it, too, must have depended on food imports, perhaps in part from Moti, half of which it ruled over. Bacan also appears to have had to import food and, according to Pires, was important in trade (*ibid.*: 218–19). Early Dutch sources suggest, however, that provisions were readily obtainable there (Meilink-Roelofsz 1962: 352 n.78). Other islands were more self-sufficient than these. Makian, for example, produced foodstuffs as well as being productive in cloves (*ibid.*: 97). Halmahera was a net food exporter, particularly of sago (*ibid.*: 98; Galvão 1544: 37). Ternate, Tidore, Makian and Kajoa had political links with specific Halmaheran villages which were used for the provision of goods and services (de Clercq 1890: 66, 72). If nineteenth century accounts are any guide (and it is doubtful if the situation had changed that much), sago and other foods were also imported from the Bacan group, Obi, Sula and also the Vogelkop of Irian Jaya (ENI 1917–35: *4*, 314–22). Similarly, sago was imported from Seram to Ambon-Lease, along traditional trading routes, perhaps facilitated by inter-island links between villages tied by relations of mutual exchange and cooperation – e.g. *pela* and *sagu ma-anu* (Bartels 1977: 140–4; Van Hoevell 1875: 54–55). At present, Ambon, with its high population density, is heavily dependent on Seramese imports, and the number of mature sago palms has been steadily decreasing. Seram is the *nusa ina* (the mother island) in an immediate subsistence sense, as well as in mythology. On Ambon itself, it seems that the smaller settlements politically tied to Hitu provided it with, in addition to cloves, wood and other products, which must have included sago.

Although part of the foodstuffs was paid in tribute, food producers also received porcelain and cloth. This is supported by comparative archaeological data, by the pattern of trade in the area for which there is documentation, and by more recent historical and ethnographic accounts. But it is likely that the bulk of the trade must have been obtained then, as now, by exchange of local produce – e.g. dried fish, metal artifacts, pottery (Ellen and Glover 1974), and other manufactured items. The ancient trade with the coast of Seram must have been important, not only in sustaining clove-producing areas during their growth but also in cushioning them following the rapid decline in the clove trade.

What we have here is a simple non-hierarchial exchange system being integrated within a wider hierarchial system: EMSUs being structured into a wider economy incorporating large numbers of both land and sea communication routes, through links based on trade in certain scarce resources. There are large numbers of petty traders compared with the volume of trade, and a large number of lower level centres provisioning a few primary centres at unfavourable terms of trade to the producers and with an inefficient movement of commodities. But rather than a static kind and rate of production, it is irregular. The conditions of growth and decline are 'abnormal'. Also the centres and radial trade routes do not eliminate local network trade (cf. Smith 1976: 7, 34–35).

On the larger islands it is possible that trade between the interior and certain coastal centres was facilitated by economic and political ties. From the seventeenth century onwards, there is evidence of the influence of the coastal rajas of Seram and Buru over interior groups, supplying them with valuables – mainly porcelain and cloth, and perhaps ironware – against sago, damar resin, rattan, cassowary and parrot feathers, plaited mats, tobacco and locally-woven cloths (e.g. Elmberg 1968: 133). It is highly probable that this pattern may go back much further and represents part of the base on which such centres as Kambello and Luhu developed. In early trade, bulk foodstuffs were unlikely to have been a significant factor, although there must have been local peripheral trade in shell and bracelets, salt, sago and such like along the coast and into the interior of the larger islands.

So local trade in foodstuffs was vital for the continued existence and growth of clove exporting centres and, as cultivation expanded, so the same areas became more dependent on the import of sago. In places it was possible to delay the reliance on external foodstuffs by more intensive appropriation of wild *Metroxylon*, its increased domestication, selection for improved varieties, the denser planting of stands and the use of other species of palm. Although this must have varied according to variety, soil moisture, topography and other factors, there appears to be a limit to which sago can be extended to dry land (Ohtsuka 1976: 8).

At the same time, the pressures on land and food resources, brought about by extensive clove cultivation, may have encouraged the development of other domesticates, for example maize and *Xanthosoma. Ipomea*, the sweet potato, was probably a late sixteenth-seventeenth century introduction to the Moluccas connected with the Portuguese (Burkill 1935: 1246; Barrau 1965: 342), although it may well have been present in Southeast Asia for much longer. Manioc, *Manihot utilissima*, which is now the most important tuber in terms of weight consumed, was not known in 1692 to Rumphius, but was in Java by 1794 (Burkill 1935: 1413). The same is true of a wide range of fruits (Van Slooten 1959: 315, 318–20). Maize appears to have been dispersed much more widely in the sixteenth and seventeenth centuries (*ibid.*: 2289). Other domesticates, such as *Amaranthus*, may have become important since the seventeenth century, but some, e.g. ground nuts and *Lagenaria*, are known for the area archaeologically from much earlier times (Glover 1977). Anyway, they seem to have displaced the yam as an important crop and, because they were more easily cultivated, have encouraged more reliance on horticulture. In places, this trend must have been accompanied by a shift from extensive swiddening to intensive dry field cultivation.

The Crisis of Dutch Monopoly and the Late Exchange Phase

During the first few decades of the seventeenth century, the Dutch

East India Company attempted to enforce a rigid monopoly on clove exports through a ruthless military campaign against other European trading countries and by virtually annihilating local Indonesian shipping in spices. Dutch activities off the coasts of Seram and Ambon led to a depletion in Asian traders. The inhabitants looked around for other places to sell their cloves, and traders sought other sources of cloves. As early as 1617 and 1618, traders were acquiring cloves from Manipa and Buru in exchange for foodstuffs, and the growth of Seram as a centre of clove production is directly related to the activities of the Dutch against foreign merchants (Meilink-Roelofsz 1962: 212).

The immediate result of the warfare between the Moluccans and the Dutch was depopulation. People left areas ravaged by the wars because of food shortages and to avoid military service; and infanticide seems to have been widely practised. The shortages were themselves a result of European presence: they had been totally dependent on clove production and imported food and now the lines of supply were interrupted.

The decline in population led to the clove trees falling into the hands of a small minority, cloves remaining unharvested and trees running wild. The population began devoting more energy to the growing of food crops and fishing, because Company rice was priced too high. In addition to selling cloves to the Dutch, taxes were paid in cloves to the ruler and his officials concerned with clove cultivation. For the same reason they preferred to turn to the growing of other food crops (*ibid.*).

In 1620, Banda was captured and destroyed by the Dutch, and so Asian trade in the Moluccas finally collapsed (*ibid.*: 220). Under the Dutch monopoly, natives were constrained to devote their time to the cultivation and gathering of the spices solely for the Company. Since Ambon yielded enough cloves to meet the combined demand of Europe and Asia, the Company concentrated all its attention on the clove villages of this island and attempted to stop trade and cultivation elsewhere, particularly Seram, in order to prevent over-production (De Klerck 1938: 256). Not only did this lead to famine and rapid depopulation, but it was also not very effective, since the inhabitants were continually finding new places in which to grow cloves (*ibid.*: 220).

Because of high transportation costs, the price of the rice or sago the Dutch sold to the natives in exchange for cloves was too expensive for the latter to buy. As a result, they began to grow rice themselves, produce their own sago, or fetch it from distant islands

(Meilink-Roelofsz 1962: 215). Such *'hongi-tochten'* offenders were punished by cutting down clove trees. This led to resistance; there were insurrections, first on Banda and then on Ambon. In 1647 there was wholesale destruction of clove trees (De Klerck 1938: 242–3). There was a revolt on Ternate. Excess sago palms and clove trees were cleared away and, in 1652, a treaty came into effect which granted the Company rights to destroy as many clove trees as it thought desirable. The work of destruction continued, and the greater part of the various islands became depopulated. The island of Buru was devastated and then, in 1654, settlements on the Hoamoal peninsula of Seram were made uninhabitable through the destruction of sago palms. The remaining population of the minor islands and parts of the west coast of Seram was transported to Ambon and Manipa. Henceforward, clove cultivation was restricted to Ambon, Haruku, Saparua and Nusalaut (*ibid.*: 256–7).

Although the Dutch policy of extirpation was ultimately successful, in a number of related ways the monopoly then went on to rebound against the Company. First, in 1656, cloves became scarce and there were no longer enough to satisfy the European and Asian market (Meilink-Roelofsz 1962: 221). It became clear that if the production of spices was not to cease altogether then the native population would have to be provided with sufficient food and cloth (*ibid.*: 227), and that the Company itself could not supply all the cloth, rice, sago and other goods that were in demand (*ibid.*: 209). But, more particularly, the Moluccas were steadily being transformed into a wilderness as the former prosperity disappeared and the settlements fell into decay (de Klerck 1938: 257). By the middle of the eighteenth century, the spice trade was declining in the Moluccas. Ternate and Tidore had lost all importance, and cloves yielded three times their cost price (*ibid.*: 424). Bickmore (1868: 154) reports that, according to official estimates, the total yield from 1675 to 1854 was 100,034 Amsterdam pounds. Although annual production was extremely variable and uncertain, this gives an average output of 558.85 Amsterdam pounds per annum. The Company attempted to diversify its interests but still maintained its control over the production and export of spices, although they had become a source of loss rather than profit. The nature of Moluccan subsistence and exchange meant that, as previously, they were cushioned against fluctuations in sales and prices. But the continued decline of the clove trade reached such an extent that even the traditional fallback subsistence position was no longer entirely effective. Ambonese and others were already having to import sago from Seram on an increasingly large scale, paying for it with the money (the Dutch had introduced this on a large scale) they were getting from decreasing clove sales. They were fully tied in a web of dependency and poverty. The deteriorating situation was partly responsible for the Uliasser revolt of 1817. Spice planting remained restricted until 1824, and the forced delivery of cloves was not finally abolished until 1864. But by then, the clove market was dominated by non-Indonesian centres and the Moluccans were compelled once again to fall-back on sago subsistence and on local and petty trade (de Klerck 1938: 2–316–7). Only later in the century did spice exports again begin to rise and become more profitable.

Banda

The small group of Banda islands south of Seram constituted an important and integral element in the trading system, ecological succession and social history we have been discussing. However, their geographical isolation, their concentration on the production of nutmeg and mace, and certain other historical factors, make them worthy of separate consideration. Moreover, they provide a critical case for the parts of the argument. Perhaps more than any other locality in the Moluccas they had long been dependent on imported food, had a high population density, an enormous output of nutmeg and mace, and were the centre of a network of trading links extending north to Ternate and Ambon, east to Kei, Aru and New Guinea (via Seram) and west to Java and Malacca. Banda was also a trading polity itself.

At the beginning of the sixteenth century, Banda appears to have had no single local ruler (Pires 1512–15: 206; Barbosa 1516: 198) and to have been constituted by small independent local polities governed by elders. However, there also appear to have been links, and even intermarriage, with Javanese, Malays and perhaps also Chinese. Islam had been recently introduced but still appeared to be limited mainly to those engaged in trade, again possibly persons with Malay and Javanese connections (*ibid.*). There is also evidence of some kind of suzerainty exercised over Banda by the sultan of Ternate, a connection related to the trading link between the two places. The Ternations were militarily superior with good oar-powered war galleys but lacked the merchant fleet of the Bandanese (Dames 1918–21: 198, n.3, following Castanheda).

By 1600 there appears to have been as much overproduction of nutmeg and mace in Banda as there was of cloves in the more northerly isles, with large quantities being destroyed (Barbosa 1518:

197). Although the Chinese were apparently not interested in nutmeg and never visited the islands (Meilink-Roelofsz 1962: 100), Malaccan, and possibly other Malay and Javanese, traders were important, exchanging Cambaya cloth of cotton and silk, coarse Sunda cloth, gongs and so on. As elsewhere in the archipelago, these items, together with porcelain, were used as valuables (Pires 1512–15: 206–7). In general, rates of exchange were exploitational (Dames 1918–21: 98–21: n.1, n.2; 202 n.1, following Conde de Ficalhio; Meilink-Roelofsz 1962: 93–5, following Pires). Goods were also exchanged for cloves. The cloves were, in turn, obtained by Bandanese merchants in the north Moluccas and Ambon, in exchange for Indian cloths (Pires 1512–15: 207; Dames 1918–21: 198 n.3, following Castanheda). The Bandanese also appear to have undertaken voyages in slave-manned galleys, as far as the New Guinea coast, Malacca and Java; in western trade, however, they were probably of small importance (Tiele 1879: 23, following Castanheda; Meilink-Roelofsz 1962: 97). So, long before the arrival of the Portuguese, Banda was a focal point of Asian trade.

The Bandanese were virtually dependent on imports for basic food and other supplies. The island of Gunung Api in the group, because of its volcano, had not nutmeg and so was a source of timber, firewood and perhaps other woodland products. Nailaka had small quantities of sago (Pires 1512–15: 206). Apart from spices, coconut palms and some fruit, there were virtually no agricultural crops on the islands. Most sago came from the Kei and Aru islands in exchange for textiles – particularly the coarse Sunda cloths – which the Bandanese had obtained from the west in exchange for spices. Rice, and other foodstuffs brought by the Javanese and Malays, were consumed by the Bandanese themselves. Because of the increasing number of nutmeg trees and the volume of trade, it was both necessary and economically sensible to import cheap food, rather than to grow it themselves. Sago was clearly of supreme importance, being occasionally used for actually making payments for nutmegs (*ibid.*: 208–9). Apart from sago, the Kei and Aru islanders also brought gold and dried parrot and bird-of-paradise plumes which met Asian demand (*ibid.*: 209). Pottery for local consumption on Banda must have come from Kei (Ellen and Glover 1974: 363). Massoi ('clove') bark, slaves and sago were also obtained from New Guinea in exchange for Sunda cloths (Elmberg 1968: 125).

The arrival of the Portuguese led to a rise in the price of a nutmeg, and throughout the sixteenth century the Banda nutmeg trade expanded. Bandanese relations with the Dutch were bad. Dutch products were disliked and stocks often remained unsold. The Bandanese wanted produce previously obtained from native traders and which the Dutch did not bring (Meilink-Roelofsz 1962: 214). Furthermore, the natives were forced to devote time to the cultivation and gathering of nutmeg for the Company. This led to famine and rapid depopulation. The Bandanese resisted, fled to the interior and were almost wiped out. Banda was finally conquered in 1621. In that year Company nutmeg sales in Europe reached 450,000 pounds of nutmeg and 180,000 pounds of mace (*ibid.*: 224). The land was given over to Dutch colonists, who were supposed to continue the cultivation of nutmegs with slave labour brought from all over Asia. Completely isolated and well-defended from the outside world, there developed a highly peculiar community, ethnically and socially. In order to maintain the monopoly, strict controls were placed on the visiting local and Asian boats, such that even small *perahus* loaded with sago were prevented from unloading (*ibid.*: 215). Instead, the inhabitants were required to live on Javanese rice. However, this was a costly business. In the end, attempts were made to revive the trade carried on by the Kei and Aru islanders, which had come to a standstill during the war of extermination, and the Company was compelled to recommend that food crops be grown instead of spices (*ibid.*: 218–19, 222). Unlike cloves, Banda nutmegs continued to be sold with profit into the eighteenth century, but the inhabitants gained a poor living and continually supplemented their incomes by smuggling (de Klerck 1938: 424). At the end of the eighteenth century, there was a series of natural disasters and the settler landowners ran into debt. In 1860, slavery was abolished; four years later, the monopoly and forced deliveries were made optional. This led to a slight revival (*ibid.*: 317).

Conclusion

I have outlined here the distinctive properties of Moluccan economies relying on the *Metroxylon* palm as their main source of starch and shown how the importance of this in the local and regional economy is intimately related to the way in which spice production and trade has developed. I have constructed a model which helps to explain, concisely and according to the known facts, the environmental and economic history of the Moluccas between 1450 and 1700. Local patterns cannot be understood except in relation to progressive ecological change and external links, as relatively independent EMSU communities were absorbed into more extensive and complex modes of production and exchange, through local polities, petty states,

Javanese and European hegemony and finally total colonial control.

It is possible to isolate three reasonably distinct trading nexuses, each focussed on a densely populated spice-producing centre, dependent on an extensive sago-producing periphery. The foci in each case are Ternate/Tidore, Ambon and Banda; the major sago areas are primarily and respectively, Halmahera, Seram and Aru-Kei-New Guinea. I have argued that first the northern Moluccas (and Banda), and then the central Moluccas, experienced the phases specified in the model, but it is clear that the chronology and duration of the phases in each area was rather different.

Sago subsistence both allowed for the rapid growth of clove production and prevented the immediate negative economic consequences that are usually associated with rapid expansion of agricultural production for exchange of crops that have no local subsistence value. The communities were able to tolerate the erratic and wide fluctuations in spice production and sales. However, it was precisely sago dependence, plus external pressure, which paradoxically maintained and increased the rate of expansion in some specific localities such that an initially adaptive pattern finally proved inadequate. On the small islands heavily involved in spice production, with the depletion of traditional subsistence resources, the system adjusted by importing food, particularly sago, from other islands not so involved in the spice trade. However, this made the communities still more dependent on clove production, although they were partly cushioned through payment in locally produced commodities rather than imported items and food. The EMSU, part of a specialised local ecosystem, was drawn into a much more specialised regional system.

My argument has been based on key historical documents, and on secondary studies and analyses. When in doubt I have gone back to some of the original sources on which the latter are based. I have relied extensively, for historical reconstruction, on accounts from more recent periods and modern parallels, and have drawn information, ideas and inspiration much more widely. In using such an eclectic assortment of materials, there is always an element of risk: over-interpreting statements made by early commentators, using sources uncritically, drawing general inferences from specific cases and placing too much faith in ethnographic analogies. All this I acknowledge. The model is one of general tendencies and patterns only. It is not a substitute for detailed archaeological, historical and ecological analysis of particular localities. Its applicability must ultimately be tested against further data.[4] However, the assumptions made are surely reasonable given what we know of Moluccan history, ethnography and environment.

But the model is of more than passing historical interest, since it may be used to explain certain aspects of contemporary change in the Moluccas. In the recent period, the production and trade in spices (and also of copra) has been steadily expanding once again, and the population of the smaller islands growing such that they are still dependent on food imports from other areas. At the same time, the geographical area of spice production has spread and even the most isolated subsistence communities have been drawn within the system. Contemporary social relations in the Moluccas cannot be understood except as part of a system which is a product of differential population growth and density, commercial expansion, political centralisation and ecological division of labour.

More generally, even this very preliminary analysis shows how it is impossible to understand the ecology, economy and social position of one natural resource in isolation, and how the changing importance of one may be closely related to the properties of another. It indicates probable processes of positive feedback and throws into relief the complex articulation of economic and ecological relations in regional systems, of which our knowledge remains relatively un-developed.

Footnotes

A number of people have commented usefully on earlier drafts of this paper, including Dieter Bartels, John Dransfield, Jim Fox, Chris van Fraassen, Tony Milner and Leontine Visser. Not all would agree with parts of my argument, but all deserve thanks, if only for bothering to disagree.

1. To avoid confusion I have used modern names wherever possible. Early names vary between authorities, localities and periods of time. *The Moluccas* (*Maluquo, Maluco*) was used by the first European voyagers to refer collectively to Ternate, Tidore, Bacan, Obi and Makian. Its use was gradually extended to include Halmahera, and then Seram, Ambon and other islands to the south (figure 3), an onomastic change consistent with the historical details discussed here. The term nowadays refers to a province of the Republic of Indonesia (see Dames 1918–21: 201, n.).

2. This fieldwork was conducted under the auspices of *Lembaga Ilmu Pengetahuan Indonesia* (the Indonesian Institue of Sciences) from 1969 to 1971, and in 1973 and 1975; and financed by the Social Science Research Council, the Central Research Fund of the University of London, the London-Cornell Scheme for East and Southeast Asia and the Hayter Travel Awards Fund.

3. For convenience all weights have been reduced to kilograms. Early sources refer to the *bahar*, a measure used widely in large trading transactions. It varied considerably between different items of merchandise and places (even

within the Moluccas). For example, Tiele (1875: 412, n.5 following
Valentijn) gives the Banda clove *bahar* as 100 *kati*, or 5.75 Amsterdam
pounds. That of the northern Moluccas is given as 200 *kati* (Galvão 1544:
287). Within the Malay area it is generally reckoned to be the equivalent
of 400 pounds or 181.4 kilograms (Yule and Burnell 1968: 47–8). Ferrand
(1921: 231) puts the northern Moluccan *bahar* of 200 *kati* at 273. 105 kgs,
and this is the equivalent I have used wherever possible.
4. The data exist in relative profusion (see e.g. de Graaf 1961; Irwin 1965;
   Coolhaas 1960, particularly pp. 47–50). The Portuguese sources in
   particular have been much underrated as material for the reconstruction
   of traditional society and economy (Boxer 1965).

References

Allen, J. 1978. Fishing for wallabies: trade as a mechanism for social interaction,
integration and elaboration on the central Papuan coast. In *The evolution of
social systems*, J. Friedman and M. Rowlands (eds.). London: Duckworth.

Barbosa, Duarte 1518. *The book of. . . completed about the year 1518*, M.
Longworth Dames (ed.) 1918–21. London: Hakluyt Society, Second Series,
44 and 49.

Barrau, J. 1958. *Subsistence agriculture in Melanesia*. Bernice P. Bishop Museum
Bull. 219: Honolulu.

——— 1959. The sago palms and other food plants of marsh dwellers in the south
Pacific Islands. *Economic Botany* 13, 151–162.

Bartels, D. 1977. *Guarding the invisible mountain: intervillage alliances, religious
syncretism and ethnic identity among Ambonese Christians and Moslems in
the Moluccas*. Cornell University Ph.D. thesis.

Bassett, D.K. 1958. English Trade in the Celebes, 1613–1667. *Journal of the
Malayan Branch of the Royal Asiatic Society* 18 (1), 1–39.

Bickmore, A.S. 1868. *Travels in the East Indian archipelago*. London: Appleton.

Bosscher, C. 1853. Statistche Aanteekeningen Omtrent de Aroe-Eilanden. *Tijd.
voor Indische Taal-, Land-en Volkenkunde* 1, 323–26.

Boxer, C.R. 1965. Some Portuguese sources for Indonesian historiography. In
An introduction to Indonesian historiography, Soedjatmoko *et al.* (eds.).
Ithaca: Cornell University Press.

Brookfield, H.C. with D. Hart 1971. *Melanesia: a geographical interpretation of
an island world*. London: Methuen.

Burkill, T.H. 1935. *A dictionary of the economic products of the Malay
peninsula*. London: Crown agents for the Colonies.

Campen, C.F.H. 1885. De landbouw op Halemahera. *Tijdschrift voor Nijverheid
en Landbouw* 29, 1–17.

Clercq, F.S.A. de 1890. *Bijdrage tot de kennis der Residentie Ternate*. Leiden:
Brill.

Cobban, J.L. 1968. *The traditional use of the forests in mainland Southeast
Asia*. Papers in International Studies, Southeast Asia Series No. 5. Athens,
Ohio: Ohio University.

Coolhaas, W.Ph. 1960. *A critical survey of studies on Dutch colonial history*.

(Koninklijk Instituut voor Taal, Land-en Volkenkunde. Bibliographical Series 4). The Hague: Nijhoff.

Dunn, F.L. 1975. *Rainforest collectors and traders: a study of resource utilization in modern and ancient Malaya.* Monographs of the Malaysian Branch of the Royal Asiatic Society. No. 5.

Ellen, R.F. 1975. Non-domesticated resources in Nuaulu ecological relations. *Soc. Sci. Inform.* **14** (5), 127–150.

—— 1978. *Nuaulu settlement and ecology: an approach to the environmental relations of an eastern Indonesian community.* Verhandelingen van het Koninklijk Instituut voor Taal-, Land- en Volkenkunde 83. The Hague: Nijhoff.

Ellen, R.F. and I. Glover 1974. Pottery manufacture and trade in the central Moluccas, Indonesia: the modern situation and the historical implications. *Man (N.S.)* **9**, 353–379.

Elmberg, J.E. 1968. *Balance and circulation: aspects of tradition and change among the Mejprat of Irian Barat.* Stockholm: Ethnographical Museum Monograph Series No. 12.

ENI 1917–35. *Encyclopaedie van Nederlandsch-Indie.* 8 vols. 's-Gravenhage, Leiden.

Ferrand, G. 1921. *Les poids, mesures et monnaies des mers du sud aux XVIe et XVIIe siècles.* Paris: Imprimerie Nationale.

Forrest, T. 1969 (1779). *A voyage to New Guinea and the Moluccas,* (edited with an introduction by D.K. Bassett). London: Oxford University Press.

Fortgens, J. 1909. Sagoe en sagoepalmen. *Bulletin van het koloniaal Museum te Haarlem* **44**, 70–104.

Fox, J.J. 1977. *Harvest of the palm: ecological change in Eastern Indonesia.* London: Harvard University Press.

Friedman, J. 1976. Marxist theory and systems of total reproduction. Part I: Negative. *Critique of Anthropology* **2** (7), 3–16.

Galvão, A. 1971 (1544). A treatise on the Moluccas, probably the preliminary version of the lost *Historia das Moluccas,* (edited, annotated and translated by H. Th. Th. M. Jacobs). Rome: Jesuit Historical Institute.

Geertz, C. 1963. *Agricultural involution, the process of ecological change in Indonesia.* Berkeley and Los Angeles: University of California Press.

Glamann, K. 1958. *Dutch-Asiatic Trade, 1620–1740.* Copenhagen and The Hague.

Glover, I.C. 1977. Prehistoric plant remains from Southeast Asia, with special reference to rice. Revised version of a paper presented at the 14th International Conference of South Asian archaeologists in Europe, Naples, July 1977.

Graaf, H.J. 1961. Aspects of Dutch historical writings on colonial activities in Southeast Asia with special reference to the indigenous peoples during the sixteenth and seventeenth centuries. In *Historians of Southeast Asia,* D.G.E. Hall (ed.). London.

Hoëvell, G.W.W. van 1875. *Ambon en meer bepaaldelijk de Oeliasers, geographisch, ethnographish, politischen historisch geschetst.* Dordrecht.

Irwin, G. 1965. Dutch historical sources. In *An introduction to Indonesian historiography*, Soedjatmoko *et al.* (eds.). Ithaca: Cornell University Press.

Johnson, D.and K. Ruddle 1976. The geographical distribution of sago-making in the old world. Paper presented at the conference 'The equatorial swamp as a resource', Kuching, Sarawak, 5–7 July.

Kantor Sensus dan Statistik Propinsi Maluku 1972. *Sensus Penduduk 1971 Propinsi Maluku.* Ambon.

—— 1974. *Maluku dalam angka 1974.* Ambon.

Klerck, E.S. de 1938. *History of the Netherlands East Indies*, 2 volumes. Rotterdam: Brusse.

Leur, J.C. van 1955. *Indonesian trade and society*. The Hague-Bandung: W. Van Hoeve.

Massalman, G. 1963. *The cradle of colonialism.* New Haven: Yale University Press.

Meilink-Roelofsz, M.A.P. 1962. *Asian trade and European influence in the Indonesian archipelago between 1500 and about 1630.* The Hague: Nijhoff.

Ohtsuka, R. 1976. The sago eaters of the Oriomo Plateau, Papua New Guinea. Paper presented at the conference 'The equatorial swamp as a resource', Kuching, Sarawak, 5–7 July.

Pires, T. 1944 (1512–15). *The Suma Oriental of Tome Pires, 1512–15*, A. Cortesao (ed.). London; Hakluyt Society, second series, 89–90.

Riedel, J.G.F. 1886. *De Sluik – en kroesharige rassen tusschen Selebes en Papua.* s'Gravenhage: Nijhoff.

Ruinen, W. 1920. *Sago palmen en hun beteekenis voor de Molukken.* Amsterdam: J.H. de Bussy.

Sahlins, M. 1972. *Stone age economics.* London: Aldine.

Slooten, D.F. van 1959. Rumphius as an economic botanist, 295–338. *Rumphius memorial volume*, H. de Wit (ed.). Baarn: Uitgeverij en Drukkerij Hollandia.

Smith, Carol A. 1976. *Regional analysis. Vol I: Economic systems.* London: Academic Press.

Spencer, J.E. 1966. *Shifting cultivation in southeastern Asia.* (University of California publications in Geography, **19**). Berkeley and Los Angeles: University of California Press.

Tiele, P.A. De Europeers in den Maleischen Archipel 1509–1623. *Bijdragen tot de Taal-, Land-en Volkenkunde.* (25 (1877), 321–420; 27 (1879), 1–19; 28 (1880), 260–340, 395–482; 29 (1881), 153–214, 332; 30 (1882), 141–242; 32 (1884), 49–118; 35 (1886), 257–355; 36 (1887), 199–307.)

Townsend, P.K. 1974. Sago production in a New Guinea economy. *Human Ecology* **2** (3), 217–236.

Wallace, A.R. 1962 (1869). *The Malay archipelago.* New York: Dover.

Yule, M. and A.C. Burnell 1968. *Hobson – Jobson: a glossary of colloquial anglo-Indian words and phrases,* W. Crooke (ed.). New York: Humanities Press.

FORAGERS AND FARMERS IN THE WESTERN TORRES STRAIT ISLANDS: AN HISTORICAL ANALYSIS OF ECONOMIC, DEMOGRAPHIC, AND SPATIAL DIFFERENTIATION[1]

DAVID R. HARRIS

Anthropologists and archaeologists interested in the evolution of economic systems, particularly those scholars who have participated in the prolonged and inconclusive debate about the origins of agriculture, have tended to focus their attention on fully agricultural societies. Working retrospectively from a knowledge of the agricultural economies that sustained historically known civilisations in Southwest Asia, Mesoamerica and elsewhere, they have probed the prehistoric past for evidence of the economic transformations that such societies underwent as they developed into larger and more complex urban polities. Much less attention has been paid to societies that did not evolve towards full dependence on agriculture although they were in close contact with crops and cultivators. As I have argued elsewhere (Harris 1977a), study of developmental pathways that did not lead to agriculture may be as fruitful in the investigation of agricultural origins as study of those that did; or, to put it more generally, examination of the non-adoption as well as the adoption of economic innovations is necessary if we are to understand how socio-economic systems change — or fail to change — through time.

There are few parts of the world where foraging and farming communities have co-existed in close proximity and where there is sufficient historical evidence to allow the development of their subsistence systems — even in the recent past — to be studied effectively One such area is Torres Strait, which long separated Australia, the southern continent of foragers, from New Guinea, the continental island of horticulturalists. The significance of the Strait as a threshold between pre-European agricultural and non-agricultural economies is enhanced by the existence of documentary records which are sufficiently detailed to allow reconstruction of the pattern of subsistence and settlement in the mid-nineteenth century when the Islanders first came into close and sustained contact with Europeans.

Because European contact was essentially maritime and occurred
relatively late, documentary data — mainly in the form of accounts
written by ships' officers, missionaries, and colonial administrators —
are unusually abundant and spatially specific. Their existence in
archives in England and Australia makes possible comparative historical
study of individual islands and island groups. Such study complements
fieldwork and allows inference from present conditions to be related
to, and sometimes tested against, historical data. Torres Strait thus
offers a unique opportunity for reconstructing the traditional subsistence systems of the island communities that occupied this transitional region between two continental areas of contrasted life style.
And such reconstruction in turn provides a foundation on which to
erect a speculative but testable hypothesis to explain why horticulture
was practised intensively on some islands but not on others.

The role of the Strait as a biogeographical filter through which
Melanesian and Australian plants and animals have moved in opposite
directions has long been recognised (Good 1947: Pl. 4; Steenis 1950;
Walker 1972), but its significance as a bridge and barrier both linking
and separating the peoples of coastal Papua and the Cape York
Peninsula has attracted less attention. Although several early European
visitors to the Strait commented on the cultural contrasts that existed
across and within it (e.g. Macgillivray 1852: 2, 1—4; Moresby 1876:
18), no systematic ethnographic study was attempted until the end of
the nineteenth century when Haddon and his associates carried out
the fieldwork that led ultimately to the publication of the six volumes
of the *Reports of the Cambridge Anthropological Expedition to Torres
Straits* (1901—35). The *Reports* still constitute the only comprehensive study of the peoples of the Strait, but in the last decade a number
of anthropologists, archaeologists, and geographers have turned their
academic attention towards this enticing arena for research (Baldwin
1976; Beckett 1972, 1977; Golson 1972; Harris 1976, 1977b; Kirk
1972; Laade 1968, 1971; Lawrie 1970; Moore 1972; Nietschmann
1977; Vanderwal 1973; White 1971; Wurm 1972).

The Strait and its Islands

At its narrowest point, between Cape York and the Papuan coast north
of Saibai Island, Torres Strait is only 150 km wide. It contains over
100 islands and islets, but only 17 of them are permanently inhabited.
They can be divided into four groups according to their location,
physical make-up, and relief: an eastern group of high islands consisting
of basic volcanic rocks; a central group of low coralline limestone
islands and reefs; a western group of high islands made up of acid

volcanic rocks; and, in the northwest adjacent to the Papuan coast, two alluvial islands (Saibai and Boigu) that rise no more than a few metres above sea level. The latter two groups together constitute the western Torres Strait Islands which are the subject of this paper.

The western Strait is underlain by the submerged Cape York-Oriomo Ridge (Jennings 1972: 34). The high islands, which consist mainly of tuffs and granites of Carboniferous age, project above sea level from the Ridge and reach maximum heights of 399 m on Moa (Banks),[2] 246 m on Muralug (Prince of Wales), 242 m on Dauan (Mt. Cornwallis), and 209 m on Badu (Mulgrave). They are grouped mainly in the southwestern Strait from the southernmost and largest, Muralug (205 km²), some 20 km from the Australian mainland, to the next largest, Moa (170 km²) and Badu (104 km²), and, lying to north and east, the much smaller islands of Mabuiag (Jervis, 8 km²) and Nagir (Mt. Ernest, 1 km²), while far to the north, only 10 km off the Papuan coast, is the little island of Dauan (3 km²) situated between the much larger, low and swampy islands of Saibai (106 km²) and Boigu (Talbot, 85 km²) (see Figure 1).

The dominant feature of the climate of the Strait is the seasonal alternation of wet (December–April) and dry (May–November) conditions. Mean annual rainfall increases northward across the Strait from 1,768 mm at Cape York to 2,063 mm at Daru (Gibbs 1971: 9; McAlpine *et al.* 1975: 22), and the dry season accordingly diminishes somewhat in length and severity towards the north. The vegetation of the islands has been much modified by human action, but open-canopy, sclerophyllous woodland occupies uncleared parts of the larger islands and – adapted as it is to the strongly seasonal climatic regime – probably represents the dominant plant community of the islands in the past. The smaller islands support mainly coastal communities of littoral thicket and woodland, as well as mangrove vegetation, strips of which also fringe the shores of the larger islands. Freshwater and brackish swamps exist on certain islands, particularly Saibai and Boigu, and small areas of closed-canopy, mainly deciduous forest clothe steep, relatively moist slopes on some of the higher islands, particularly Dauan. There are no areas of closed-canopy evergreen rain forest. Grasslands occur in the interior of the larger islands, but, on Moa and Badu at least, there is evidence that these are the result of the burning and clearance of woodlands for subsistence purposes.

The pattern of physical, climatic, and vegetational variations between and within islands gives rise to significant differences in available resources. The deepest and most fertile soils occur on the

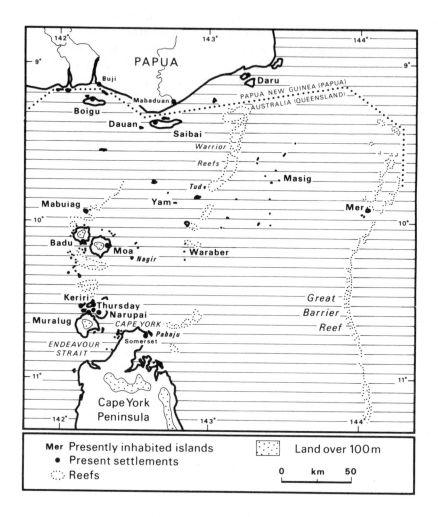

Fig. 1. The Torres Strait Islands (only islands mentioned in the text are named).

weathered basic rocks of the eastern islands and this is reflected in
more luxuriant growth of vegetation, both wild and cultivated.
Around the eastern islands the sea is deeper than it is farther west;
fish and shellfish are abundant, but marine turtles and dugongs ('sea
cows', *Dugong dugong*), which together comprise the Islanders'
principal traditional source of protein and fat, are less plentiful than
in the shallower seas of the central and especially the western islands.

On the low sandy and muddy central islands there is little deep, cultivable soil, vegetation is more sparse, and fresh water is scarce, although the reefs provide rich fishing grounds. The acid rocks of the high western islands yield thinner, less fertile soils than the eastern volcanics, and there are extensive outcrops of massive granitic boulders, but the greater size of the western islands partly compensates for their less cultivable terrain. Fringing reefs surround the islands, and large platform reefs have built up between them from Muralug to Mabuiag. These reefs are inhabited by a great variety of fish, crustaceans, and molluscs; and green turtles (*Chelonia mydas*) and dugongs feed off the marine monocotyledons or 'seagrasses' that grow abundantly in the shallower waters. The low alluvial islands of Saibai and Boigu are not surrounded by growing reefs but by extensive, shallow mudbanks and mangrove swamps which allow only canoes and dinghies to reach the few landing points. The muddiness of the sea along the Papuan coast restricts but does not prohibit the growth of seagrasses so that dugongs and turtles are somewhat less plentiful here than farther south in the Strait, and the absence of reefs similarly reduces the variety of fish available. The extensive swamps that occupy much of the interior of the two islands limit the amount of cultivable land, but they do support duck and other edible waterfowl.

Supplies of wild terrestrial plant and animal foods are meagre on all the islands compared with the resources available on the mainlands to north and south of the Strait, but they are rather more diverse and abundant on the larger western islands with more varied relief such as Muralug, Narupai (Horn), Moa, and Badu, than on the smaller and physically more uniform islands. There are no native land mammals (except bats), although in pre-European times the Islanders possessed domesticated dogs, and pigs were introduced to some of the islands from New Guinea (Haddon 1912: 137, 152). Terrestrial animal foods were limited to such minor items as birds, lizards (*Varanus* spp.), snakes, frogs, and beetles. Birds were probably the most important source of such food. They included ducks, seabirds and their eggs, and the Torres Strait pigeon (*Myristicivora spilorrhoa*) which migrates south across the Strait in the dry season. The mound-building scrub fowl (*Megapodius freycinet*) occurred on the larger western islands (Gill 1876: 314) where its eggs and flesh were eaten. There is no doubt, however, that in the past most of the animal food consumed by the Islanders came — in the form of fish, shellfish, turtles, and dugongs — from the marine ecosystems of inter-tidal foreshore, reef, and shallow sea.

Islanders and Europeans

Since 1879 the Torres Strait Islands have been part of the State of Queensland (Farnfield 1974: 68). Today the total population of the 17 islands with permanent settlements slightly exceeds 6,000. Some 3,000 people live on Thursday Island, the administrative and commercial centre of the Strait, and approximately 400 on the nearby islands of Narupai, Keriri (Hammond), and Muralug. The remaining 13 inhabited islands are reserved for indigenous Torres Strait Islanders and each has its own elected council. In 1974, when I carried out fieldwork in the six western Reserve Islands, the estimated populations of the western, and of the central and eastern, Reserve Islands were respectively 1,640 and 1,203. These communities consist almost exclusively of Torres Strait Islanders, whereas the population of Thursday, Narupai, Keriri, and Muralug islands is more heterogeneous and includes many people of European origin.

Opportunities for gathering first-hand information about traditional economic activities are now largely limited to the Reserve Islands. Even there such activities have been greatly curtailed in recent decades by increases in social service payments to, and opportunities for cash employment among, the Islanders (Duncan 1974). This decline has affected horticulture and wild plant-food procurement more than it has fishing, turtling, and dugong hunting; and it has been more intense in the southwestern Reserve Islands close to Thursday Island than in the more remote northwestern islands off the Papuan coast. Fortunately the data derived from fieldwork can be greatly enriched by reference not only to Haddon's monumental study but also to several earlier nineteenth-century accounts of Torres Strait, particularly those that derive from the surveying voyages of H.M.S. Rattlesnake (Macgillivray 1852; Moore 1974) and H.M.S. Basilisk (Moresby 1876), and from the reports and letters of the earliest missionaries (Gill 1876; McFarlane 1888; Murray 1876 and the London Missionary Society archive).

During the first two centuries of intermittent European exploration of the Strait, following Torres' passage through it in 1606, contact with the Islanders was too slight and fleeting to induce significant social change. But after the first British settlements in Australia were established, European ships began to pass more frequently through the Strait and the Islanders began to acquire iron knives and other novel goods in considerable quantities. The navigational dangers of the reef-strewn Strait came to be seen as a threat to communication with the colony of New South Wales, and the Admiralty dispatched first the Fly in 1842 and then the

Rattlesnake in 1846 to chart its waters thoroughly.

The voyages of the Fly and the Rattlesnake initiated a new phase of painstaking maritime survey and ethnographic description. The geologist Jukes sailed as naturalist on board the Fly and subsequently published the first general account of the physical geography and natural history of the Strait, which includes many ethnographic observations of the eastern islanders (Jukes 1847), as does the journal of Sweatman who sailed with the Fly on board H.M.S. Bramble (Allen and Corris 1977). The survey was continued by the Rattlesnake, to which Macgillivray was attached as naturalist and Brierly as artist. Both wrote detailed accounts of the voyage (Macgillivray 1852; Moore 1974) and performed a notable ethnographic service by recording the first-hand knowledge of life on Muralug acquired by a Scottish woman from Sydney, Barbara Thompson. She was the only survivor of a shipwreck off Cape York and she lived for four years with the native inhabitants of Muralug before being rescued by the Rattlesnake.

By 1846 (Macgillivray 1852: 1, 308) Europeans had begun to exploit the Strait for bêche de mer or trepang (edible 'sea slugs', *Holothuria* spp.), for which a market had long existed among the Chinese communities of Southeast Asia. The bêche de mer fishermen soon acquired an evil reputation as ruthless exploiters of the local population, and the situation deteriorated further after the pearlshell (*Pinctada maxima*) fishery was started at the Warrior Reefs in 1868 (Farnfield 1975: 72–4; Beckett 1977). The navigational dangers, growing commercial importance, and increasing lawlessness of the Strait had already persuaded the Queensland Government to found a settlement at Cape York. This was established on Pabaju (Albany Island) in 1863, transferred to Somerset on the mainland in 1864, and finally moved to Thursday Island in 1877 (Farnfield 1974: 47–7, 1975; Jardine 1866). With an administrative presence established in the Strait and legal authority extended to all the islands, the worst excesses of the fishermen were at last curbed.

A desire to mitigate the cruelty of the pearlshell and bêche de mer fishermen was one reason for the establishment of Christian missions in the Strait in the last three decades of the nineteenth century. The initiative came from the Congregational London Missionary Society which had already established missions in the Southwest Pacific. The pioneer missionaries Samuel McFarlane and A.W. Murray first visited the Strait in 1871, bringing with them converted Pacific Islanders to act as teachers (McFarlane 1888; Murray 1876). Mer (Murray) became the headquarters of missionary activity and in the western islands evangelism was largely left to the Pacific Islanders. The accounts of missionary life left by McFarlane, Murray, and other missionaries (e.g. Gill 1876) provide useful ethnographic information, particularly on the late-nineteenth century distribution of population, but from about 1890 the activities of the London Missionary Society in the region shifted increasingly to Papua. In 1914 the Society handed over all its lands and buildings in the Strait to the Church of England (Bayton 1965; White 1917), and a new generation of European missionaries took up their self-appointed task.

Evangelical endeavour had profound social and economic effects on the island communities among whom the missionaries worked. It hastened the modification of traditional culture which had begun during the earlier phases of maritime contact. It encouraged the rejection of 'pagan' beliefs and practices and fostered commercial enterprise. In the terminology of the time, Christianity 'brought the Lights' and the Islanders began gradually to consign their traditional mores to what they came to call the 'darkness time'. When Haddon first visited the Islands in 1888 (Haddon 1890), the populations of several were already overtly Christianised, but on this visit and during his main expedition in 1898 he was nevertheless able to record many aspects of traditional culture.

After 1900 the process of modernisation gathered pace. In 1904 a commercial company, Papuan Industries Limited, was founded by a former missionary who opened a company store on Badu, one object of which was to assist the western Islanders to plant coconuts and to produce copra (Haddon 1935: 17). In 1930 the Queensland Government bought the company store on Badu, and by that time government teachers had begun to take over the education of the Islanders from the missionaries. Since the Second World War the Queensland Government has increased its efforts to develop the Islands, spurred on by the Australian Federal Government (Duncan 1974; Fisk 1975) which began to pay more attention to the Strait as the date agreed for the independence of Papua New Guinea (1975) approached. Today the people of the 17 inhabited islands regard themselves as Queenslanders, but despite their long exposure to European ideas and institutions they have, in the Reserve Islands, retained elements of their indigenous culture. It is still possible to learn much about traditional life from the Islanders, and this information, combined with judicious use of the European record, allows us to elicit, in outline at least, the mid-nineteenth century pattern of population, economy, and society.

Pre-European Society

As European knowledge of the Islands increased through the nineteenth century it became apparent that their inhabitants differed considerably in physique, language, and culture, not only from the Aborigines of the Australian mainland but also, among themselves, from one island group to another. Physically the Islanders were seen to resemble Papuans more closely than Aborigines, although those living on Muralug and other islands close to Cape York were said to look somewhat like Aborigines. Linguistically there was a basic division between the eastern islands where the people spoke a Papuan language called Miriam, which belongs to the Eastern Trans-Fly language family, and the western and central islands where Mabuiag, which is structurally related to Australian Aboriginal languages, was spoken (Wurm 1972: 364–9). These languages, which are still used in the Islands, are themselves distinct from the Australian languages that were spoken by Aborigines of the Cape York Peninsula and from the Papuan languages that are spoken west of the Fly River.

Before sustained European contact disrupted traditional culture, the total population of the Islands probably numbered between 4,000 and 5,000 people divided into some five or six kin-based communities, each of which occupied two or more islands. Some communities were largely sedentary while others were seasonally mobile, but each exercised customary rights over the resources of a given area and also participated in an inter-island trade network. This network extended to the Papuan coast as well as to the northern Peninsula and goods were traded from island to island in double-outrigger sailing canoes. Raids and marital exchanges also gave rise to inter-island and island-mainland movement between communities.

The basic units of the pre-European social system were exogamous villages and hamlets in the sedentary and horticultural eastern islands and exogamous patriclans or bands among the more mobile, less horticultural populations of the mid-western and southwestern islands; while the people of the northwestern islands, who depended substantially on horticulture as well as on wild-food procurement, lived in permanently established villages which were divided into clan wards (Beckett 1972: 320–5). Throughout the Islands each clan exploited its own stretch of coast and also formed part of a larger territorial district. On Mabuiag, for example, there were, before the missionaries brought the formerly dispersed population into one village, at least seven clans grouped into four districts (Haddon 1904: 163, 172, 266–7). The clans had totemic as well as territorial significance, each being associated with one or more of the 36 totems (31 animals, 2 plants, and 3 inanimate objects) that Haddon recorded (1904: 154—7) in the western islands. The districts had head men invested with religious authority who tended to be the leaders of the clans with the most important totems but who were not chiefs or 'big men' in a political or economic sense. The basic resource-exploiting unit was thus the clan or band which consisted of a number of family or hearth groups and which itself belonged to the larger territorial groupings of district and community.

Pre-European Economy and Population in the Western Islands

Historical evidence demonstrates that in the western islands in the mid-nineteenth century there was a broad north-south transition across the Strait from reliance on gardening to reliance on foraging. But the economic emphasis on wild-food procurement or horticulture also varied in relation to differences in the position, size, and physical resources of individual islands and island groups. Thus horticulture was well established on the three northern islands, less important on Badu and especially Moa, and practised only intermittently with very few crops on Muralug; contained within this north-south gradient was a secondary pattern of more intensive horticulture on the smaller inhabited islands of Dauan, Mabuiag, and Nagir. These variations in the traditional economies of the western islands can be correlated with inter-island differences in resource availability, but they also relate to mid-nineteenth century patterns of community organisation, socio-economic exchange, and population density: patterns that require closer examination before economic, demographic and spatial differentiation can be analysed at the scale of individual island groups.

Insular Communities and Exchange

In his first ethnographic report on the peoples of the western Strait Haddon records (1890: 301) that they themselves recognised four island groups, the inhabitants of which were 'distinctly allied': the Muralug group and Moa; Badu and Mabuiag; Boigu, Dauan, and Saibai; and the remaining western and central islands, principally Nagir, Yam (Turtle-backed), Tud or Tutu (Warrior), and Masig (Yorke). The alliances that linked these communities were based primarily on inter-marriage, raiding, and the reciprocal exchange of food and other resources. They were not strongly reinforced by linguistic differentiation, although island-group dialects did exist.

Barbara Thompson's testimony, recorded by Brierly on the Rattlesnake in 1849, is a rich source of information on inter-island socio-economic relationships. It is clear from her comments, and from the later accounts of trade and warfare provided by Haddon, that the nature of these relationships varied with the inter-island and island-mainland distances involved. At the local, intra-community scale contact was frequent and informal; at the intermediate, inter-community scale it was less frequent and more formalised; and at the regional island-mainland scale it took the form of systematised trade.

One trade item – the dugout outrigger sailing canoe – eclipsed all others in importance in island-mainland exchange. The hulls were cut and hollowed out at the mouth of the Fly River, fitted with a single outrigger and floated along the Papuan coast to Saibai where they were decorated, re-fitted with two outriggers and gunwale, and thence traded south via Dauan, Mabuiag, and Badu to Muralug; they also reached Muralug by an alternative route via Tud and Nagir (see Figure 1). Haddon describes this trade (1904: 296–7) and gives details of the methods by which Islanders ordered and paid for canoes, using as currency shells, dugong harpoons, and, from Muralug, pieces of iron obtained from European wrecks. Trade between the islands and the Australian mainland was much less intensive and less significant economically than the New Guinea trade. The only Islanders who had close contact with the Aborigines at the northern tip of the Cape York Peninsula were the inhabitants of Muralug and adjacent islands. Barbara Thompson comments that 'The mainland Blacks are constantly backwards and forwards, but they bring nothing but spears sometimes, which they make better than our people. They just come for what they can get to eat amongst us' (Moore 1974: A67); and Haddon suggests (1904: 295) that 'throwing sticks and javelins' were 'probably the only imports' and that they may have 'found their way to Mabuiag'.

A long-distance trade network thus spanned the Strait, linking the western islands with the mainlands to north and south. Products manufactured from resources – particularly shells, feathers and timber – obtained mainly or exclusively either in insular or in mainland environments were regularly exchanged, the systematisation of the trade increasing with the rarity and socio-economic importance of the object traded. It is unlikely that food entered the long-distance trade network in substantial quantities. Dried turtle (and possibly dugong) meat was eaten on long-distance voyages, but it is doubtful whether any other staple foods were regularly traded between the islands and the mainlands.

On the other hand, much inter-island movement had as its object the reciprocal exchange of, and access to, food resources, and it took

place both between and within island communities. Clear evidence of
this is provided by Barbara Thompson's statement that the people of
Nagir brought yams, sugar cane, sago, and other plant products to
Muralug (Moore 1974: A66). Significantly, she adds that

> The Kulkalaga [of Nagir] bring their things over as presents for our people,
> not by way of exchange. They eat our people's *kotis* [wild yams] . Some-
> times our people may take something across as Peaqui did the dried fish,
> and the Kulkalagas gave him a canoe. When the Kulkalagas went away this
> year, they gave Peaqui and Manu each a canoe. . . . This canoe was given
> to Manu by a young man of the Kulkalagas who wanted to have Yesu . . .
> for a wife (Moore 1974: A66–7).

This passage aptly describes a relationship of generalised reciprocity
whereby near neighbours (between whom marriage is sanctioned)
exchange goods as gifts rather than through a more formalised system
of barter such as regulated the cross-Strait canoe trade. The quotation
and other references by Barbara Thompson to the links between the
peoples of Muralug and Nagir (e.g. Moore 1974: A86) imply that, in
the mid-nineteenth century at least, the latter island formed part of
the Muralug-Moa community and that it was less closely linked than
Haddon supposed with such small central islands as Tud and Masig.

Barbara Thompson also describes a two-month visit by the friendly
Kulkalagas that evidently involved the temporary migration to Muralug
of the entire population of Nagir, which she estimates at 'at least 300'
(Moore 1974: A65–6). It occurred towards the end of the dry season
when wild yams were still available on Muralug, and it is clear from
the account that on this occasion the generalised reciprocity existing
between the two populations took the form of an exchange of
manufactured goods from Nagir (bamboo knives and containers for
water and tobacco, coconut water containers and fibre lines for
securing sucker fish, bows and arrows, mats and plaited waistbands,
neck ornaments, and processed sago) for access to a dry-season supply
of staple food on Muralug. That this was not an isolated or exceptional
event is clear from Barabara Thompson's comment that the Kulkalagas
'Come to Moralug every year, staying sometimes many months, as it
might be from the end of one season to the beginning or part of the
next. They like the *koti* that grows on our island' (Moore 1974: A24).
There is no equivalently detailed account of exchange between
Muralug and the other large island in its community – Moa – but
Barbara Thompson confirms Haddon's association of the two islands
and mentions the existence of marriage ties between them.

The socio-economic relationships that bound together other inter-
island communities in the western Strait cannot be demonstrated in

such detail, but there is little doubt that a pattern of exchange between large and small islands also characterised communities north of the Muralug group. In his first, brief description of Badu, Haddon remarks (1890: 407) that

> the inhabitants are very closely allied in speech and by marriage with the Mabuiag people, but they have very little communication with the natives of Moa — though the latter island is separated from Badu by a shallow channel which averages only a mile and a half across.

The subsistence economies of Badu and Mabuiag complemented each other in the sense that the larger island was better provided with wild terrestrial food resources and the smaller was better placed for access to the richest fishing and dugong-hunting grounds. Horticulture was practised on both islands but appears to have been more intensive on Mabuiag than Badu, before its general decline following the rise of the pearlshell and bêche de mer industries. In a relationship of generalised reciprocal exchange between the two islands there was probably a tendency for wild terrestrial resources from Badu to be exchanged for horticultural and marine produce from Mabuiag. A similar relationship of generalised interdependence probably linked the three northern islands of Dauan, Saibai, and Boigu, with Dauan being the most intensively cultivated and the two larger islands providing wild acquatic and terrestrial resources in greater abundance. However, this inter-island community differed from that of Badu-Mabuiag in having close social and economic connections with Papuan coastal villages across the narrow channels that separate the three islands from the mainland.

Population Density

The historical sources allow reconstruction, in outline at least, of the mid-nineteenth century patterns of community organisation and exchange, but they provide less adequate data for the analysis of inter-island variations in population density. The problem of inadequate demographic data is further complicated by the mobility of many populations in the western Strait, especially those of the smaller islands such as Nagir who periodically migrated *en masse* to other islands and whose enumeration by early European observers was therefore particularly difficult and prone to error.

   The only numerical estimates that date from the mid-nineteenth century derive from the records of the Rattlesnake expediation and relate to Nagir and Muralug. As we have seen, in her description of the Kulkalagas' visit to Muralug in 1849 Barbara Thompson guessed that the population of Nagir amounted to 'at least 300'. On the other hand Macgillivray, who visited Nagir in December 1849 when most of its people were absent visiting Waraber (Sue), estimated (1852: 2, 42) the population of the entire 'Kulkalega tribe' at '100 souls'. Barbara Thompson comments that the Kulkalagas came in 'at least a dozen canoes' and Macgillivray gives (1852: 2, 16) the capacity of a Muralug canoe as 25 people. The canoes of Nagir and Muralug were of the same type, but since the visitors brought large quantities of goods with them, it is unlikely that each canoe contained more than 15–20 individuals. Therefore a total population for Nagir of approximately 200, half-way between the two estimates, seems likely. Barbara Thompson gives the population of Murulug as 'not above 50' and adds that 'There have been many deaths since I first landed there. There are not many people now' (Moore 1974: A11, A7). In view of her long residence on the island, this estimate is unlikely to be grossly inaccurate and her reference to 'many deaths' probably reflects the impact of such diseases as measles and influenza introduced from European ships following the then customary route through Endeavour Strait.

The populations of the other western islands remain unknown until the records of the London Missionary Society begin to yield some fragmentary data from the early 1870s. The earliest estimate of the populations of Mabuiag and Moa date from 1872 and 1875 respectively. On his second missionary voyage to the Strait in 1872, Murray gave the population of Mabuiag as '300 or more' (LMS I: 2/8–11); and in 1875 McFarlane reported that Moa

> is a sickly island, and during the last few years about half the population have been removed by the pearlshellers, and by disease. The [missionary] teacher has not been able to find more than about 250 people on the island, and they are living in the interior, having fled from the coast to avoid the pearlshellers who are exceedingly anxious to get them for divers' (LMS I: 10/1).

The 'teacher' had already left Badu 'on account of there being but few natives there' (LMS I: 10/2) and the implication is clear that the populations of both islands had already been substantially reduced — probably halved — as a result of European contact. Most of the reduction is likely to have taken place in the early 1870s following the start of the pearlshell fishery at the Warrior Reefs in 1868. The fact that the fishery had its headquarters first at Somerset and later at Thursday Island implies that all the populations of the Western Islands from Mabuiag southward were subject to the rapacious recruiting demands of the pearling ships, demands that escalated rapidly through the 1870s (Yonge 1930: 165–6). Mabuiag became

one of the main bases of the fishery (Moresby 1876: 130) and by
1876 McFarlane was reporting a new estimate of 170 for the population of Moa: 'and these are afraid to live near the beach on account
of the shellers' (LMS I: 16/2). If McFarlane's two estimates are
accepted, we find that the population of Moa was reduced by about
80, or 32%, in the fourteen months between April 1875 and August
1876.

There are regrettably few references in the early missionary records
to the numbers of people living on the three northwestern islands.
Murray and McFarlane comment in the report on their first voyage
to the Strait that the 'population of Dauan is very small, but there
seems to be a constant intercourse between it, and Saibai, the
population of which appears large', furthermore 'The two islands . . .
have close relations . . . with the natives of the [Papuan] coast'
(LMS I: 1/52–3). Moresby, who visited Dauan in 1873, two years
after the missionaries first landed there, believed (1876: 133) that
the village was 'only occasionally occupied, as the natives live on
Saibai', and in 1881 it was said of Dauan that 'there are not 100
people on the island altogether, and three-fourths of the people that
are there belong to Boigo and stay there occassionally (*sic*)' (LMS II:
22A/16). Moresby also visited Saibai in 1873 and stated (1876: 133)
that it is 'well populated, and the principal village contains about 600
inhabitants', but by 1884 there were, according to Strachan (1888:
24) who spent several days on the island, about 130 people living in
the one village. This apparent decline probably reflects in part the
impact of disease and recruiting, but the frequent movement that
evidently took place between the three islands and to and from the
Papuan coast may also have given rise to errors in the early estimates.
Boigu was less frequently visited by Europeans than any other of the
inhabited western islands and there is very little information on its
late-nineteenth century population. Strachan gives (1888: 131) its
population prior to his visit in 1885 as 350. This figure refers to a
time before the village was attacked by the notorious Tugeri warriors
from western Papua who periodically raided the northwestern islands
and the Papuan coastal villages to which the islands were allied.
Refering to the Tugeri, McFarlane claims (1888: 106) that Boigu 'has
been almost depopulated by these cannibals'.

In attempting to estimate the pre-contact populations of the
western islands, allowance must clearly be made for the depopulating
effects of introduced diseases and recruitment by the pearl shellers.
In the more remote northwestern Strait missionaries reached the
islands before the pearl shellers and protected the islanders success-

fully from recruitment (Gill 1876: 209; LMS I: 16/4), although less effectively from disease (Strachan 1888: 73, 75). The estimates of 600 and 350 for the populations of Saibai and Boigu relate sufficiently closely to the time of effective contact in 1871 to be regarded as approximations to pre-contact populations, but Moresby's statement that the people of Dauan lived on Saibai suggests that his estimate of 600 for Saibai should be taken to include Dauan. The reference to Dauan's population being very small in 1871 probably reflects — as its context suggests — the temporary absence of most of its people on Boigu, Saibai, or visiting the Papuan coastal villages. Thus the population of Dauan appears to have been only semi-permanent and seldom to have exceeded about 100. On this basis the pre-European population of the three northwestern islands can be estimated at approximately 950 (see Table I). By 1890, when an enumeration of island populations was recorded (LMS III: 34/13), this total had been reduced to some 400.

It is more difficult to arrive at well-founded estimates of the pre-European populations of the other western islands. The figures of 300 and 250 for Mabuiag and Moa in 1872 and 1875 respectively cannot represent pre-contact populations because by then disease and

TABLE I

Estimated Pre-European (c. 1840) Populations and Population Densities of the Western Torres Strait Islands

	Area	Pop.	Density /km²	Coast km	Density cst./km
Muralug	204.9	100	0.5	83	1.2
Other islands of the Muralug group	88.9	150	1.7	79	1.9
Muralug group	293.8	250	0.8	162	1.5
Nagir	1.0	200	200.0	4	50.0
Moa	170.5	500	2.9	52	9.6
Muralug-Moa-Nagir Community	*465.3*	*950*	*2.0*	*218*	*4.4*
Badu	104.4	670	6.4	47	14.2
Mabuiag	8.3	300	36.1	10	30.0
Badu-Mabuiag Community	*112.7*	*970*	*8.6*	*57*	*17.0*
Dauan	3.0	100	33.3	7	14.3
Saibai	106.4	500	4.7	61	8.2
Saibai-Dauan	109.4	600	5.5	68	8.8
Boigu	85.1	350	4.1	62	5.6
Saibai-Dauan-Boigu Community	*194.5*	*950*	*4.9*	*130*	*7.3*
Western Torres Strait Islands	*772.5*	*2870*	*3.7*	*405*	*7.1*

recruiting had already taken some toll of the Islanders. If McFarlane's statement that Moa's population was halved in a few years is taken literally, a pre-contact population of 500 can be assumed, although this does not allow for any demographic changes that may have resulted from intermittent European contact before the rise of the pearlshell fishery. If it is assumed that Mabuiag's population had similarly been halved by 1872, a pre-contact total there of some 600 can be postulated. There is no comparable early estimate for Badu, but as has already been suggested, it had probably lost half its population by 1875. Its pre-European population can be inferred if it is assumed that the rate of demographic decline between 1875 and 1890 (when its population was 124) was the same on Badu as on Moa. From Table I it can be calculated that Moa's population was reduced by a factor of 2.7 in that period. If the population of Badu in 1890 is multiplied by 2.7 we get a computed 1875 population of 335 which can then be doubled to give an estimated pre-contact population of 670.

Uncertainties also arise when one attempts to assess the pre-European populations of Nagir and Muralug, even though very early estimates are available. It is unlikely that by 1849 the population of Nagir had been significantly reduced by contact with Europeans, and as has already been argued, the Kulkalaga 'tribe', which was based on Nagir but migrated to other islands, probably numbered about 200 in the mid-nineteenth century. Muralug's early exposure to European influences makes estimation of its population particularly difficult. However, by analogy with the rapid halving of population that apparently took place when Moa was first exposed to European raids and diseases, Muralug's pre-contact population can best be estimated by doubling Barbara Thompson's figure of 50. A total of 100 is probably the right order of magnitude for Muralug alone, but it does not make allowance for the other southwestern islands. The only hint of the populations of these islands are comments by Murray that in 1872 there were a few people on Narupai and a considerable number of Keriri (LMS I: 2/9). Without more precise evidence we can only guess that the islands other than Muralug were probably occupied by a further 150 people, giving an estimated pre-contact population of 250 for the Muralug group as a whole.

Having estimated the pre-contact populations of the western islands, it is possible to draw some general conclusions about community size and inter-island variations in population density. The estimated total population of 2,870 gives an overall density of 3.7

people per km² of island area, or an average of 7.1 people per km of coast (see Table I). If individual islands are examined it is seen that crude population density varies from 0.5/km² on the largest island, Muralug, to as high as 200/km² on the smallest, Nagir. The three smallest islands for which population estimates are available — Nagir, Dauan, and Mabuiag — had the highest densities, and it was on them that horticulture was most highly developed. Conversely, the Muralug group, where cultivation was scarcely practised, exhibits the lowest densities, with an overall figure of less than one person per km². These contrasts are examined more closely in the final section on the subsistence systems of island groups, but it is first necessary to draw attention to the remarkable consistency in community size that Table I reveals. Despite the extreme variations in population totals and densities that occur from island to island, the three inter-island communities that functioned as socio-economic units have populations and densities of the same magnitude. They range in size only from 950 to 970 and in density from 2.0 to 8.6 per km². Each community includes one of the three most densely peopled and intensively cultivated islands, whose population interacted socially and economically with the occupants of the larger, less horticultural island(s) in the community. Thus Nagir was associated with Muralug and Moa, Mabuiag with Badu, and Dauan with Saibai and Boigu in a spatial and seasonal pattern of subsistence that articulated with the exchange network that spanned the Strait.

Insular Subsistence Systems

The three pre-European communities of the western Strait provide the framework within which to reconstruct the mid-nineteenth century pattern of resource use and population; and this in its turn clarifies the role of the Strait as a socio-economic threshold between Australia and New Guinea. Although an overall north-south gradient from more to less horticultural economies did exist, it is only when analysis of economic, demographic, and spatial differentiation is carried out at the scale of individual islands and inter-island communities that the complexity of the traditional pattern is revealed.

(a) The Muralug-Moa-Nagir Community (see Figure 2c). Horticulture appears to have been less well established on Muralug and adjacent islands than on any other of the inhabited western islands. Macgillivray commented (1852: 2, 25) that 'at the Prince of Wales Islands the cleared spots are few in number, and of small extent, — nor [do they] naturally produce either the cocoa-nut or bamboo, or

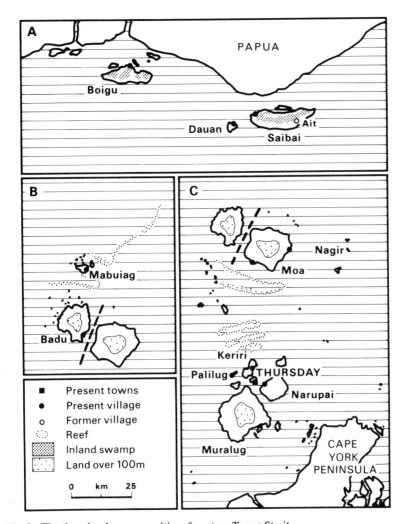

Fig. 2. The three insular communities of western Torres Strait.

is the culture of the banana attempted'. He goes on to describe (1852: 2, 26) how yam gardens were prepared by clearing and burning vegetation at the end of the dry season and states that yams were 'the most important article of vegetable food' because they last 'nearly throughout the dry season', but it is apparent from the more detailed references to yam utilisation in Barbara Thompson's testimony that most of the yams that were eaten were gathered wild. Her description of yam planting on Muralug demonstrates that it was

very casual and resorted to mainly to ensure a supply in case there
were insufficient wild yams (Moore 1974: A43). It is also clear that,
with the exception of sugar cane (Moore 1974: A43), the plants that
the people of Nagir brought to Muralug were not cultivated there.
Subsistence on Muralug and adjacent islands depended mainly on
wild-food procurement. The staple sources of protein, fat, and
carbohydrate were evidently marine turtles, dugongs, fish,
crustaceans, molluscs, wild yams, wild fruits, and mangrove shoots
(the germinating embryos of *Bruguiera gymnorhiza* and possibly
other species of the family Rhizophoraceae).

The sparse human population was seasonally mobile (Haddon
1890: 428). During the dry season, when wild yams provided the
main supply of carbohydrate, the people foraged in the hilly interior
as well as along the coast, obtaining roots, fruits, and shellfish, and
making occasional canoe voyages to fish and to visit nearby islands
as well as the mainland west of Cape York. At this time of year
family or hearth groups travelled independently, sleeping in the
open or in temporary tent-shaped shelters (Haddon 1912: 94–5). The
last two months of the dry season (October–November), when the
dominant winds shift from southeast to northwest heralding the
approach of the rains, were the main turtle-hunting time when
breeding pairs floating on the surface of the sea could be harpooned
relatively easily from canoes. At this time too, everyone congregated
at a chosen site to build a communal dwelling where they then
remained throughout the wet season (Moore 1974: A17–8). The
communal house was located on the southern, sheltered coast of
Muralug 'on the only creek on the whole island where they can take
the canoes up at high water' and where they were near the mangroves
'on which they mainly subsist during the worst part of the [wet]
season' (Moore 1974: A90, A93).

Subsistence on Muralug was thus closely adjusted to the strongly
seasonal climatic regime. The people were semi-sedentary, in the
sense that each wet season they returned to and remained at the
same site and each dry season they moved their camps at intervals
while foraging and fishing on land and sea. Although they received
horticultural produce from Nagir, in exchange for granting the
Kulkalagas a share in the harvest of wild foods — especially yams —
on Muralug, they did not commit themselves to horticulture.
Cultivation was casual and intermittent, a minor activity supplement-
ing wild-food procurement. Customary rights were, however, exercised
over cultivable land and other resources. Macgillivray states (1852: 2,
28) that 'According to Gi'om [Barbara Thompson], there are laws

regulating the ownership of every inch of ground on Muralug and the neighbouring possessions of the Kowraregas'. Even allowing for the possible imposition by Macgillivray and Barbara Thompson of European concepts of land tenure on to a less formal indigenous system of customary rights, the evidence does suggest that access to resources was controlled at the level of family or hearth groups and that all available territory was at least notionally 'owned'.

The social and demographic dimensions of the subsistence system cannot be reconstructed in detail but they can be inferred in outline. By 1888, when Haddon first visited Muralug, the pre-European social system had largely disintegrated, but by relating his accounts (1904: 173–4) of totemism on Muralug and on Tud and Yam it is possible to infer that the population of Muralug was formerly divided into two clans or bands. If this situation prevailed prior to effective European contact, and if the estimate of 100 for the pre-contact population is approximately correct, then each band may have consisted of some 50 people. For Muralug alone, this implies an average of 41.5 km of coast per band, in the context of an overall pre-European population density of 0.5/km² or 1.2 people per km of coast (see Table I). However, these ratios ignore the fact that the people of Muralug had some reciprocal access, with other island populations in the Muralug-Moa-Nagir community, to resources in the much larger area of land and sea exploited by the community as a whole. If the Muralug group alone is considered the population densities work out at 0.8/km² or 1.5 people/coastal km, and for the whole community they rise to 2.0/km² or 4.4/coastal km (Table I).

The latter figures reflect in particular the effect of the high population densities indicated for the tiny horticultural island of Nagir (200/km² or 50/coastal km). There is no doubt that there were areas of intensive cultivation on Nagir (Macgillivray 1852: 2, 3), but the computed ratios for the island exaggerate the real situation there because, as we have seen, the Kulkalagas regularly visited Muralug for social and economic purposes as well as other islands such as Waraber to fish and forage. The ratios for Moa (2.9/km² or 9.6/coastal km) are much closer to those of Muralug than to those of Nagir. This accords with expectation, because, although horticulture was more firmly established on Moa than on Muralug, the people of both islands were basically fishers and foragers rather than cultivators.

There is no direct evidence of the number of clans or bands into which the pre-European populations of Nagir and Moa were divided, but if the figure of 50 inferred for Muralug is assumed to apply to the rest of the community we can postulate the existence of ten bands on Moa, four on Nagir, and three on the other islands of the Muralug group. This gives a total of 19 bands for the community as a whole and implies an average of 11.5 km of coast per band. An indication that the inference of 10 bands on Moa may be approximately correct can be gleaned from the sketch map of the island compiled by Wilkin (in Haddon 1904: 8). It shows the location of twelve coastal places which may be the sites of wet-season clan settlements, although unfortunately neither the number nor the distribution of clans on Moa is discussed. No doubt the number and size of bands varied somewhat from generation to generation, but there is abundant evidence that population growth was regulated not only by intra-community marriage and migration but also by infanticide and other direct and indirect methods of control such as abortion and prolonged suckling (Haddon 1890: 359; Moore 1974: A11, A17). Indeed Macgillivray states explicitly (1852: 2, 11) that 'The population of Muralug is kept always about the same numerical standard by the small number of births, and the occasional practice of infanticide'. This evidence, coupled with that for the elaboration of customary rights over land and other resources as well as for socio-economic reciprocity between islands, suggests that — before it was disrupted by European influences — the Muralug-Moa-Nagir community functioned as an integrated subsistence system in which a long-term balance was maintained between resource availability and population density.

(b) The Badu-Mabuiag Community (see Figure 2b). The complementarity that existed within the southwestern community between horticulture on Nagir and wild-food procurement on the larger islands of Muralug and Moa is paralleled in the mid-western community by the economic interdependence of Badu and Mabuiag. Although the contrasts in island size and economic differentiation between Badu and Mabuiag are not so extreme as those between Nagir and Muralug-Moa, they are of the same type. Horticulture was practised on both Mabuiag and Badu, but it was evidently more intensive on the smaller island. According to Wilkin (in Haddon 1904: 284) the gardens of Mabuiag 'were once second only to the sea as a source of subsistence' and the abandoned remains of quite elaborate, partly terraced field systems can still be observed on the island. I did not see any comparable vestiges of former field systems on Badu, although there is no doubt that a range of crops was traditionally cultivated there by swidden and other techniques. This is clear from Landtman's notes on Badu

horticulture recorded by Haddon (1912: 148—9) which describe the cultivation there 'from time immemorial' of taro, yams, sweet potatoes, sugar cane, bananas, and coconuts. However the existence on Badu of a folktale that records the attempts of the legendary figure Yawar to teach people how to raise yams in mounds (Haddon 1904: 36—7) may refer back to a time when the island was largely non-horticultural. Certainly wild terrestrial foods were more varied and abundant on Badu, which is twelve times larger than Mabuiag, whereas on Mabuiag wild-food procurement focused more on the exploitation of fish, turtles, and other marine resources, especially the dugong; indeed, according to Haddon (1890: 351) Mabuiag was 'the headquarters of the fishery of this sirenian'.

There is little specific information about seasonal movement and settlement in the Badu-Mabuiag community. Between Haddon's visits to Badu in 1888 and 1898 rectangular, thatched houses of a type introduced by Pacific Islanders were generally adopted, but in 1888 he had observed (1935: 63) 'huts consisting of little more than two sloping walls meeting like a roof, evidently the indigenous type'. These resembled the temporary shelters used on Muralug in the dry season and they probably had a similar function on Badu. There is no definite reference to communal wet-season huts on Badu, but Haddon describes (1912: 97) huts on Mabuiag with as many as ten doors and says that the dwellings of Mabuiag and Badu were the same. It is likely that on Badu individuals and family groups spent considerable time foraging independently during the dry season and came together as clans or bands in the wet season when they occupied larger communal houses. Year-round settlements may also have been maintained close to the main areas of cultivation. On Wilkin's sketch map of Badu (in Haddon 1904: 8) thirteen coastal places are shown, which probably represent the sites of clan-focused, wet-season or permanent settlements.

It is difficult to establish the number of clans or bands on Badu before European influences altered the social system, but there is reason to believe that it was about twelve. In Haddon's discussion of the clans of Mabuiag and Badu (1904: 162—71) seven are associated primarily with specific locations on Mabuiag, four with the islet of Pulu between the two islands, and three definitely with Badu. In addition, five possible former clans are mentioned (1904: 169—70), traces of which were detected in the genealogies of individuals living on Mabuiag in 1898. They may have been territorially associated with either island or, at different times, with both. The islet of Pulu was a sacred place (Haddon 1904: 3—5) which was not permanently inhabited, and the clans associated with it were linked to Badu as well as to Mabuiag. Two of them may have originated in Badu and subsequently become established on Mabuiag, one probably moved from Mabuiag to Badu, and the territorial origin of the fourth is unknown. This suggests that Pulu — which Haddon refers to (1912: 97) as 'the common meeting place for the inhabitants of both islands' — fulfilled a regulatory function for intra-community migration by providing clans in process of territorial adjustment with the necessary spatial identity. And that clan fission and migration did take place, within as well as between the islands, can be inferred from the recorded division of one of the Mabuiag clans into two sections living on opposite sides of the island (Haddon 1904: 164).

Before the missionaries induced the people 'who formerly lived scattered over the island' to congregate in one village, the seven clans that Haddon located specifically on Mabuiag occupied distinct areas, mainly along the northern and southeastern coasts (Haddon 1904: 7, 163, 172). If this number of clans is related to the inferred pre-contact population of 300 we obtain a mean clan or band size of 43 people and an average of 1.4 km of coast per band; while the population densities for the island as a whole are $36/km^2$ or 30/coastal km (Table I). If we further assume that the remaining twelve clans of which there is some record were primarily associated with Badu — and the existence of thirteen coastal places on Wilkin's sketch map of Badu strengthens this supposition — we obtain a mean clan size of 56 people and a ratio of 3.9 km of coast per band, with average population densities of $6.4/km^2$ or 14/coastal km (Table I). The demographic contrast between the two islands that these figures demonstrate accords with the differences in the extent to which their subsistence economies were committed to horticulture, fishing, and foraging. It also demonstrates the rationale of their co-existence as one socio-economic community. At the community scale mean clan size was 51, with a ratio of 3 km of coast per band, and average population densities were $8.6/km^2$ or 17/coastal km (Table I). The mean clan or band size of 51 corresponds remarkably closely to the figure inferred for Muralug, despite the fact that the more productive horticultural-fishing-foraging economy of the Badu-Mabuiag community supported much higher densities of population than the mainly foraging and fishing economy of the Muralug-Moa-Nagir community. Population growth in the Badu-Mabuiag community was evidently regulated not only by intra-community marriage and migration but also by such methods as abortion, infanticide, and late marriage (Haddon 1904: 197–8, 247), so that, as in the southwestern islands, the community as a whole maintained a durable balance between available resources and population.

(c) The Dauan-Saibai-Boigu Community (see Figure 2a). In the three northwestern islands, as in the mid-western and southwestern communities, there was an inverse correlation between island size and intensity of cultivation. Saibai and Boigu are respectively thirty-five and twenty-eight times larger than Dauan, and it was on the small, high island — which receives more rain and has a more dependable supply of fresh water than either of the large, low islands — that horticulture was most intensive. When Moresby visited Dauan in January 1873 he observed (1876: 132) on the 'north-eastern side' of the island where 'the village and native mission station are placed . . . some fine patches of grassy land, well supplied with fresh water, and a richly cultivated valley, producing taro and melons'. Today the same valley is still favoured for gardens and on the alluvial soils that encircle the island's mountainous core many areas of former mound cultivation can be seen beneath a cover of grasses and shrubs. Grassy areas now occupy about one-third of the island. Some are the result of the decline of horticulture in recent decades, but others are swidden gardens being fallowed, just as, no doubt, were the 'fine patches of grassy land' that Moresby saw near the village in 1873.

On Saibai and Boigu there is no comparable evidence for such extensive areas in relation to island size being cultivated now or in the past. On Saibai gardens are restricted to the higher areas that rise a metre or two above the level of the coastal and interior swamps that occupy some two-thirds of the island. The gardens are concentrated in the west and are connected to the present village on the northwest coast by a network of paths. The swampy interior, with its seasonally flooded depressions, grassy flats, and stands of screw pine (*Pandanus* spp.), supports fish and wild fowl, as do the coastal mangrove swamps and creeks. Moresby commented on the 'large brackish lagoon within, which abounds with curlew, wild duck, and other wild fowl' and concluded (1876: 133) that the people of Saibai 'have plenty of vegetables and fish, of pigs, in which the island abounds, and a supply of turtle and the flesh of the dugong, which is very good eating and tastes rather like veal'. With the exception of the feral pigs, which have now been killed off to protect gardens from the damage they inflict by rooting, Moresby's comment on the mixed diet of the islanders remains applicable today. A similar mixed economy based on limited horticulture and the procurement of wild terrestrial and especially aquatic resources

supported the population of Boigu, the only significant difference
being that swampy land is even more extensive than on Saibai and
cultivation accordingly more restricted. Wild plants made a limited
contribution to the food supply on both islands. They included wild
yams, mangrove shoots, and wild fruits, especially the wongai plum
(*Manilkara kauki*) which was an important dry-season food throughout the western islands.

Prior to the arrival of the first missionaries in 1871 the people of
the northwestern islands were already living in villages, although
groups evidently moved quite frequently between islands and the
population as a whole had close socio-economic ties with the
villages of the nearby Papuan coast. The earliest European visitors
were impressed by the quality of the houses, which were more
elaborate than the temporary shelters and wet-season huts of the
other western islanders. Moresby described (1876: 133) the houses
on Saibai as 'well sized, and two stories high – the latter a peculiarity
not elsewhere seen by us'. Murray (1876: 456), who was in the first
missionary party to land on Dauan and Saibai, found the people
living in houses built mainly of bamboo 'on stakes eight or ten feet
high . . . the roofs . . . thatched, and the sides . . . enclosed with the
pandanus leaf'; and Haddon (1912: 100 and Pl. xx) photographed a
small 'pile-dwelling' on Saibai which he believed to be the 'original
type' of house in the northwestern islands.

In 1871 the population of the three islands was apparently distributed among four villages: one each on the sites of the present
villages on Dauan and Boigu and two on Saibai. One of the latter
was called Saibai; it occupied the same site as the present village on a
narrow coastal strip of higher land backed inland by swamp and in
an optimal location for access to and from Dauan. The other was
called Ait; before its inhabitants were moved to Saibai village by the
missionaries it occupied an area of higher ground at the southern edge
of the great swamp in the eastern interior of the island (Laade 1971:
xxiii). It appears to have been the only sedentary settlement in the
western islands that was sited away from the coast. It could be reached
by canoe when the swamp was flooded, as well as on foot, and its
location probably reflects the importance of access to the resources
of the interior swampland as well as the value of a refuge from the
Tugeri raiders of western Papua.

Although settlement in the northwestern islands was sedentary, in
the sense that permanently established villages existed, villagers did
move regularly within and between islands. Laade notes (1971: xxiv)
that during the main planting period in the late dry season, before

the rains, people used to camp at their garden places which were named and sometimes misleadingly referred to as villages. Movement between islands evidently occurred frequently, and, at least from Dauan, sometimes involved temporary total evacuation. It has already been pointed out that Moresby believed the people of Dauan to live mainly on Saibai, and that in 1881 another observer stated that three-quarters of Dauan's population belonged to Boigu where they occasionally stayed. Such comments suggest that Dauan may have functioned largely as a horticultural 'out-station' for the larger, less cultivable islands and that during the main wet-season period of crop growth, between planting and harvesting, most or all of the people may have lived with their relatives on Saibai and/or Boigu. The fact that Moresby visited Dauan in the wet-season month of January and concluded that the island was only occasionally occupied supports this inference. Canoe voyages between the islands and to and from the Papuan coastal villages for social visits and the exchange of produce took place throughout the year; and Laade records (1971: xxiii) that when the few waterholes on Saibai and Boigu dried up in November and December the islanders obtained fresh water from Dauan and the Papuan villages.

The number of clans into which the pre-European population of the northwestern islands was divided is uncertain. Haddon only worked briefly on Saibai in 1898 and he states (1904: 171, 177—9) that there were then five totemic clans on the island divided into two exogamous groups, the combined population of which was 353; he also describes (1904: 174—5) how the village formerly consisted of two rows of houses facing each other which were occupied by the two groups and which were themselves divided into five clan clusters. The plan of the village thus mirrored the social structure of the population. Haddon does not discuss the clan system of Boigu or Dauan, although he does remark (1904: 178) that the people 'of Dauan as well as those of . . . Boigu . . . habitually intermarried with those of Saibai, and now they all live, or practically so, in Saibai'. Laade, who worked in the northwestern islands in 1963—65, states (1971: xxv) that the population was divided into seven clans grouped into two moieties; he refers to but does not explain the existence of 'slight variations on the three islands' and his discussion (1971: xix—xxi) of the relationship between clans and legends implies that most if not all of the clans were represented on each of the islands. While there may have been 21 clans in all, it is most improbable that there were 7 on each island. There may well have been 7 clans on Boigu, but in view of the small size of Dauan and the very close social links that existed between it and the much larger island of Saibai, it is unlikely that 7 clans existed as territorial units on each of the latter islands. A more acceptable estimate is that there were 12 clans divided unevenly between Saibai and Dauan. This estimate is based partly on the contrasted size of the two islands and partly on Haddon's record (1904: 181) of the existence on Saibai — in addition to the 5 main clan totems — of 7 subsidiary ones which he suggests may indicate the extinction or absorption of former clans. If this interpretation is correct, the 12 clans were probably distributed un- evenly between the two islands with perhaps 10 on Saibai and 2 on Dauan.

Whatever the precise inter-island distribution of clans may have been, the principal conclusion that emerges from this analysis is that the total number of clans in the community corresponds to the in- ferred totals in the mid-western and southwestern communities. If the clan total of 19 is divided into the estimated pre-contact population and length of coastline of the northwestern islands we obtain a mean clan size of 50 and a ratio of 6.8 km of coast per clan. For Saibai-Dauan and Boigu separately mean clan size is likewise 50, with ratios respectively of 5.7 km and 8.8 km of coast per clan. Comparable population densities are $4.9/km^2$ and 7.3/coastal km for the community as a whole; $5.5/km^2$ and 8.8/coastal km for Saibai- Dauan; and $4.1/km^2$ and 5.6/coastal km for Boigu; while, if Saibai and Dauan are treated — unrealistically — as separate socio-economic entities, an extreme demographic contrast between them appears: $4.7/km^2$ and 8.2/coastal km on Saibai, $33.3/km^2$ and 14.3/coastal km on Dauan (Table I).

Such a contrast corresponds closely to that between Badu and Mabuiag, and, as in the mid-western community, it demonstrates the complementarity of a more intensive horticultural economy on the smaller island and a more mixed fishing-foraging-cultivating economy on the larger. It is also evident that in the northwestern islands population growth was regulated both by intra-community marriage and by infanticide and other direct methods of control; indeed Moresby concludes his description of Saibai with a reference (1876: 134) to the general prevalence (by implication there and elsewhere in Torres Strait) of 'the crime of infanticide'. Thus in the northwestern islands, as well as in the mid-western and southwestern communities, the evidence suggests that the subsistence system resulted in a long- term balance being maintained between available resources and population.

Conclusion

Socio-economic Regularities

The principal if unexpected result of this enquiry into the mid-nineteenth century subsistence systems of western Torres Strait has been to demonstrate the existence there of three insular communities with almost equal populations, linked by an exchange network in manufactured goods but dependent for their basic subsistence on the complementary exploitation of wild foods and cultivated crops in the physically contrasted islands of which each community consisted. In each community a small, high island, relatively well endowed with cultivable soil and dependable supplies of fresh water (Nagir, Mabuiag, Dauan), was intensively cultivated and part of its horticultural produce was consumed by the populations of the larger islands in the community whose economies focused more on wild-food procurement.

Although this complementary economic relationship characterised each community, there were significant inter-community differences in population densities which related to variations in the productivity of community economies. Thus while the average densities inferred for the western islands as a whole are 3.7/km² and 7.1/coastal km, those for the southwestern community fall 46 and 38 per cent below the average and those for the northwestern and mid-western communities rise respectively 32 and 3 per cent, and 132 and 139 per cent, above it. These demographic contrasts reflect the relatively high productivity of the horticultural-fishing-foraging economy of the Badu-Mabuiag community, the relatively low productivity of the mainly foraging and fishing economy of the Muralug-Moa-Nagir community, and the intermediate productivity of the more balanced economy of the Dauan-Saibai-Boigu community. They also underline the fact that the importance of horticulture in insular economies was less a function of distance from the Papuan coast − from where, it is assumed, crops and horticultural techniques were initially introduced − than of island size, resource endowment, and position relative to other islands.

The investigation has also revealed striking regularities, not only in the magnitude of community populations but also in the number and size of their constituent clans or bands. Each community appears to have contained some 19 clans, each of which consisted of approximately 50 people. There is evidence of clan fission, fusion, and extinction, which was no doubt greatly accentuated by the impact of European navigators, traders, missionaries, and administrators, but the existence of such pronounced regularities in the size of the main social units of community and clan suggests the operation of self-correcting subsistence systems of resource use, mobility, and population control. How these systems may have functioned in the mid-nineteenth century has been outlined, as far as space and the evidence permits, in the previous section.

Trade and the Development of Horticulture

This synchronic analysis provides a foundation for diachronic study of the origins and evolution of subsistence systems through earlier, pre-European times. Such study depends ultimately on the acquisition of archaeological and palaeoecological data in the islands and on the mainlands to the north and south, but the preliminary historical analysis presented here suggests a hypothesis that could explain the differential development of horticulture in the islands and could have relevance to the wider debate about the emergence of agriculture in other parts of the world.

It is clear from the nineteenth-century pattern of insular economies that horticulture was most highly developed on one small island in each community, but it cannot be convincingly argued that these three islands were inherently better suited to cultivation, in terms of their soil and water resources, than all the other islands in the western Strait. Evidently horticulture on Dauan, Mabuiag, and Nagir complemented wild-food procurement on the larger islands, but this does not explain why those particular three islands became foci of horticultural production. The clue to this question may be found in the strategic positions that they occupied in the long-distance trade network of the Strait.

The trade in canoes, which originated at the mouth of the Fly River, passed from Dauan in the northwestern community to Mabuiag in the mid-western community and thence to Badu and Muralug, while another route reached the southwestern community via Tud and Nagir. Thus each of the three small, 'horticultural' islands was situated at a critical point in the inter-community trade network where canoes and other goods passed from one community to another. The populations — permanently or temporarily resident — on the three islands are likely to have been disproportionately large in relation to island size — as the demographic data indicate — because they supported and were in turn supported by the trade network. Evidence has already been cited that the Kulkalagas of Nagir provided the people of Muralug not only with canoes but also with manu-factured goods and horticultural produce, most of which probably came from Nagir itself. Local specialisation in horticulture and the acquisition and manufacture of such trade items as shell ornaments, stone axes, and dugong harpoons can also be inferred — although not demonstrated as unequivocally — for Mabuiag and Dauan. For example, Landtman reports (1927: 34) that the coastal Papuans north of the Strait obtained their stone axes and clubs from the Islanders who fashioned them from stones brought up from the sea bed: an activity that is likely to have been best developed on Dauan as the northernmost high island consisting of volcanic rocks.

It may be hypothesised, therefore, that as both the demand for canoes and other exotic goods, and the exchange network that provided them, developed, so horticulture and the manufacture of domestic goods were stimulated on those islands the strategic location of which best fitted them to participate in long-distance trade. Indeed, that there was a direct as well as an indirect connection between horticulture and trade is suggested by Landtman's comment (1927: 215) that 'Custom requires every seller of a canoe to provide it with food to be used on the journey'. Such a hypothesis could help to explain why intensive horticulture appears to have developed only in the three small islands of Dauan, Mabuiag, and Nagir, although crops and techniques of cultivation were known throughout the western Strait.

Studies of more complex societies in Melanesia and Indonesia (for example Allen 1977; Harding 1967; Hughes 1973; Strathern 1971: 93–114; Ellen and Fox, this volume) have shown that participation in a developing exchange network can stimulate social and economic specialisation. It is possible that this process may underlie the adoption or non-adoption, and the intensification, of agriculture in some less complex (especially insular?) societies. The 'trade-horticulture hypothesis' that I am tentatively advancing for the western Torres Strait Islands should be testable if diachronic data can be obtained through archaeological and palaeoecological investigation, because it predicts that evidence for specialisation in horticulture and local manufacture will be clearest, and probably earliest, on the three small islands, As yet little more than archaeological reconnaissance has been undertaken in the islands (Vanderwal 1973), but they invite — and would probably reward — detailed investigation.

An historical analysis of the eastern islands, comparable to the one attempted here for the western Strait, could also yield interesting comparative data because there, too, small horticultural islands were linked by an exchange network to the western Strait and the Papuan coast. Beguiling prospects of further enquiry thus exist, but for the moment we must rest content with a synchronic analysis that demonstrates some of the socio-economic regularities that existed in the nineteenth-century island world of the western Strait: a world that buffered but did not divide the continent of Australia from the continental island of New Guinea.

Footnotes

1. I thank the Councillors and people of each of the Reserve Islands that I visited for welcoming me and allowing me to carry out fieldwork; also John Buchanan of the Department of Aboriginal and Islander Advancement at Thursday Island for helping to arrange my itinerary. I gratefully acknowledge the financial support for the fieldwork that I received from the Leverhulme Foundation, the Australian Institute of Aboriginal Studies, and the Central Research Fund of the University of London. The archival work was supported by a grant from the British Social Science Research Council and invaluable assistance was provided by Barbara Field who acted as Research Assistant.
2. With the exception of the administrative centre of Thursday Island (pre-European Waiben), indigenous island names are used throughout this paper, although the European names, which are given in parentheses (except for Saibai which lacks one), are used by most authors for the islands of the Muralug or Prince of Wales group.

References

Allen, J. 1977. Fishing for wallabies: trade as a mechanism for social interaction, integration and elaboration on the Central Papuan coast. In *The evolution of social systems*, J. Friedman and M. Rowlands (eds.). London: Duckworth.

Allen, J. and Peter Corris (eds.) 1977. *The journal of John Sweatman*. St. Lucia: University of Queensland Press.

Baldwin, James A. 1976. Torres Strait: barrier to agricultural diffusion. *Anthrop. J. of Canada* **14**, 10–17.

Bayton, John 1965. *Cross over Carpentaria*. Brisbane: Smith & Paterson.

Beckett, J.R. 1972. The Torres Strait Islanders. In *Bridge and barrier: the natural and cultural history of Torres Strait*. (Research School of Pacific Studies, Dept. of Biogeography and Geomorphology Publication BG/3), D. Walker (ed.). Canberra: Australian National Univ.

——— 1977. The Torres Strait Islanders and the pearling industry: a case of internal colonialism. *Aboriginal History* **1**.

Duncan, Helen 1974. *Socio-economic conditions in the Torres Strait: a survey of four Reserve Islands*. (Research School of Pacific Studies, Department of Economics) Canberra: Australian Nationa Univ.

Farnfield, Jean 1974. The moving frontier: Queensland and the Torres Strait. *Lectures on North Queensland History*. Townsville: James Cook Univ., 63–72.

——— 1975. Shipwrecks and pearl shells: Somerset, Cape York, 1864–1877. *Lectures on North Queensland History, Second Series*. Townsville: James

Cook Univ., 67—76.

Fisk, E.K. 1975. *Policy options in the Torres Strait.* (Research School of Pacific Studies, Department of Economics) Canberra: Australian National Univ.

Gibbs, W.J. 1971. *Climatic survey northern region 16 — Queensland.* Canberra: Bureau of Meteorology, Department of the Interior, Commonwealth of Australia.

Gill, W. Wyatt 1876. *Life in the southern isles.* London: The Religious Tract Society.

Golson, J. 1972. Land connections, sea barriers and the relationship of Australian and New Guinea prehistory. In *Bridge and barrier: the natural and cultural history of Torres Strait.* (Research School of Pacific Studies, Department of Biogeography and Geomorphology Publication BG/3), D. Walker (ed.). Canberra: Australian National Univ.

Good, Ronald 1947. *The geography of the flowering plants.* London: Longmans.

Haddon, A.C. 1890. The ethnography of the western tribe of Torres Straits. *J. R. Anthrop. Inst.* **19**, 297—442.

———— 1901—35. *Reports of the Cambridge anthropological expedition to Torres Straits.* (Vol. I General Ethnography 1935, Vol. II Physiology and Psychology 1901 and 1903, Vol. III Linguistics 1907, Vol. IV Arts and Crafts 1912, Vol. V Sociology, Magic and Religion of the Western Islanders 1904, Vol. VI Sociology, Magic and Religion of the Eastern Islanders 1908). Cambridge: University Press.

Harding, T.G. 1967. *Voyagers of the Vitiaz Straits.* Seattle: Washington University Press.

Harris, D.R. 1976. Aboriginal use of plant foods in the Cape York Peninsula and Torres Strait Islands. *Australian Institute of Aboriginal Studies Newsletter, New Series* **6**, 21—2.

———— 1977a. Alternative pathways toward agriculture. In *Origins of agriculture,* C.A. Reed (ed.). The Hague: Mouton.

———— 1977b. Subsistence strategies across Torres Strait. In *Sunda and Sahul. Prehistoric studies in Southeast Asia, Melanesia and Australia,* J. Allen, J. Golson and R. Jones (eds.). London: Academic Press.

Hughes, Ian 1973. Stone-age trade in the New Guinea inland. In *The Pacific in transition,* H. Brookfield (ed.). London: Arnold.

Jardine, J. 1866. Description of the neighbourhood of Somerset, Cape York, Australia. *J. R. Geog. Soc.* **36**, 76—85.

Jennings, J.N. 1972. Some attributes of Torres Strait. In *Bridge and barrier: the natural and cultural history of Torres Strait.* (Research School of Pacific Studies, Department of Biogeography and Geomorphology Publication BG/3), D. Walker (ed.). Canberra: Australian National Univ.

Jukes, J. Beete. 1847. *Narrative of the surveying voyage of H.M.S. Fly* (Two volumes). London: Boone.

Kirk, R.L. 1972. Torres Strait — channel or barrier to human gene flow? In *Bridge and barrier: the natural and cultural history of Torres Strait.* (Research School of Pacific Studies, Department of Biogeography and Geomorphology Publication BG/3), D. Walker (ed.). Canberra: Australian National Univ.

Laade, Wolfgang 1968. The Torres Strait Islanders' own traditions about their origin. *Ethnos* **33**, 141–58.

—— 1971. *Oral traditions and written documents on the history and ethnography of the northern Torres Strait Islands, Saibai-Dauan-Boigu. Volume 1. Adi – myths, legends, fairy tales.* Wiesbaden: Steiner.

Landtman, G. 1927. *The Kiwai Papuans of British New Guinea.* London: Macmillan.

Lawrie, Margaret 1970. *Myths and legends of Torres Strait.* St. Lucia: University of Queensland Press.

LMS. The archives of the Council for World Mission (incorporating the London Missionary Society). Papua journals, Boxes I & II. London: Library, School of Oriental & African Studies, University of London.

Macgillivray, John 1852. *Narrative of the voyage of H.M.S. Rattlesnake* (Two volumes). London: Boone.

McAlpine, J.R. *et al.* 1975. Climatic tables for Papua New Guinea. *Division of Land Use Research Technical Paper No. 37.* Melbourne: Commonwealth Scientific and Industrial Research Organization.

McFarlane, S. 1888. *Among the cannibals of New Guinea.* London: London Missionary Society.

Moore, D.R. 1972. Cape York Aborigines and Islanders of western Torres Strait. In *Bridge and barrier: the natural and cultural history of Torres Strait.* (Research School of Pacific Studies, Department of Biogeography and Geomorphology Publication BG/3), D. Walker (ed.). Canberra: Australian National Univ.

—— 1974. *The Australian-Papuan frontier at Cape York, II. Documentary material* (Typescript). Canberra: Library, Australian Institute of Aboriginal Studies.

Moresby, John 1876. *Discoveries and surveys in New Guinea and the D'Entrecasteaux Islands.* London: Murray.

Murray, A.W. 1876. *Forty years' mission work in Polynesia and New Guinea.* London: Nisbet.

Nietschmann, Bernard 1977. The wind caller. *Natural History* **86**, 10–16.

Steenis, C.G.G.J. van. 1950. The delimitation of Malaysia and its main plant geographical divisions. In *Flora Malesiana Series 1, Spermatophyta Vol. 1,* C.G.G.J. van Steenis (ed.). Djakarta: Noordhoff.

Strachan, John 1888. *Explorations and adventures in New Guinea.* London: Low, Marston, Searle & Rivington.

Strathern, A. 1971. *The rope of Moka. Big-men and ceremonial exchange in Mount Hagen, New Guinea.* Cambridge: University Press.

Vanderwal, R. 1973. The Torres Strait: prehistory and beyond. *Occasional Papers* **2**, 157–94. St. Lucia: Anthropology Museum, University of Queensland.

Walker, D. (ed.) 1972. *Bridge and barrier: the natural and cultural history of Torres Strait* (Research School of Pacific Studies, Dept. of Biogeography & Geomorphology Publication BG/3). Canberra: Australian National Univ.

White, Gilbert 1917. *Round about the Torres Straits: a record of Australian*

Church missions. London: Society for the Promotion of Christian Knowledge.

White, J. Peter 1971. New Guinea and Australian prehistory: the 'Neolithic problem'. In *Aboriginal man and environment in Australia*, D.J. Mulvaney and J. Golson (eds.). Canberra: Australian National University Press.

Wurm, S.A. 1972. Torres Strait — a linguistic barrier? In *Bridge and barrier: the natural and cultural history of Torres Strait* (Research School of Pacific Studies, Dept. of Biogeography & Geomorphology Publication BG/3), D. Walker (ed.). Canberra: Australian National Univ.

Yonge, C.M. 1930. *A year on the Great Barrier Reef.* London: Putnam.

TERRITORIAL ADAPTATIONS AMONG DESERT HUNTER-GATHERERS: THE !KUNG AND AUSTRALIANS COMPARED

NICOLAS PETERSON

A cultural materialist research strategy sees social organisation and ideology as the outcome of the interaction of techno-environmental and demographic process. Such a perspective might lead one to expect that hunter-gatherers living in broadly similar environments such as the Kalahari, Central Australia and the Great Basin would display some similarity in their cultural adaptations. Until recently, the belief in the widespread distribution of the patrilineal band provided this similarity. However since the 'Man the Hunter' conference in 1966, it is clear that few ethnographers have seen a true patrilineal band (but see Williams 1968: 129 and 1974), and that it has rarely existed anywhere. The traditional law-and-order view of hunter-gatherer territorial organisation has now been replaced with an emphasis on seasonal variation in group size and fluidity of group composition (see Lee 1976a: 95–96).

The arguments for fluidity are persuasive and well rehearsed. It allows for the levelling out of demographic variance, adjustment of group size to resources and the resolution of conflict by fission (Lee and DeVore 1968: 8) in a way that the patrilineal band encapsulated in its territory never could. But the switch from ideological rigidity to ecological flexibility leaves a problem. Has the cultural materialist research strategy adopted by Lee (1972a), with its heavy emphasis on behaviour[1] to the neglect of ideology, advanced our understanding at the expense of the real problem posed by the research strategy: namely why the differences in ideology are so marked when the techno-environmental conditions are similar. Whereas formerly the ethnography was interpreted as providing evidence for a patrilineal local organisation and consequently a patrilineal ideology among the San (see Steward 1936 and Service 1965) and the Australians (see Radcliffe-Brown 1930) now neither, it is generally agreed, have patrilineal bands and only the Australians a patrilineal ideology. A once substantial similarity is now a marked difference.

Is this change more than a reflection of new anthropological blinkers and of the new cultural environments in which contemporary hunter-gatherers lead out their lives or does it represent an advance in understanding? I shall argue that the over-emphasis on fluidity and the neglect of !Kung ideology has obscured the marked similarity that remains. I will begin by considering the pattern of !Kung territorial organisation and then compare it with desert Australian patterns to elucidate the nature and significance of the similarities and differences.

!Kung Territorial Organisation

Until recently the information on San territorial organisation in general and on the !Kung in particular has been limited, leaving ample room for divergent interpretations of the underlying organisational principles. Since Marshall's work in the 1950s and the Harvard expeditions in the 1960s there has been a dramatic increase in the quality and quantity of information but only limited clarification. Part of the problem has been the tendency to ignore !Kung ideology or to treat it as epiphenomenal to an understanding of territorial organisation[2] and part the polemical context of much of the discussion.

The slender evidence used by Steward to encompass the San among the patrilineal band societies was drawn from Schapera's review of the literature on the Khoisan people in 1930. On the basis of his review Schapera had concluded that there was evidence for patrilocal residence among the Naron, Heikum and !Kung but nowhere except among the Heikum was there any indication of clans (1930: 81−86). Thirty years later Marshall (1960, see 1976) confirmed that there were neither patrilineal clans nor patrilocal bands, many men spending long periods living with their wife's family, but indicated that there were headmen in each band who passed their position on to their sons (1960: 339, but see 1976: 191−195 where she has modified her views on headmanship). However it was not until Lee and other members of the Kalahari Research Group started publishing their findings that the degree of the discrepancy between the patrilineal band model and the !Kung territorial organisation began to be emphasised.

The extent of visiting, movement and variability in group composition has led Lee to characterise !Kung society as one, 'in which everyone could live wherever, and with whom, he or she pleased' (1972a: 178 recontextualised in 1976: 77) and Yellen to speak of the 'seemingly constant and random movement of individuals and families' (1977: 133). Yet it is clear that the !Kung living arrangements do have a stable basis, as Lee himself says elsewhere (1972c: 351), and it is possible for Yellen to write that the:

> population is divided into groups or bands of varying size, each based on a single permanent water source and each utilizing a particular area which includes that water point. . . Each of these areas has a sufficient mix and amount of food and most other raw materials to support the group over the course of the year' (nd: 17).

The apparent inconsistency arises because the first set of statements were concerned only with band composition. To think of a band simply in terms of an aggregate of people[3] is to focus attention on the diversity of individual behaviours and motivations and the variability in group composition, to the neglect of all else. A definition of band must include a spatial dimension if it is to be useful (see Helm 1969: 213; Turnbull 1968 and Woodburn 1968). Band membership is a cultural construct and not self-evident from behavioural observation. Band members are not always to be found in a single camp within their territory: some people may be away hunting, others off trading or visiting with the avowed intention of returning, or the band may split into small groups exploiting dispersed resources within its territory (e.g. Marshall 1965: 248 and 1976: 180; Schapera 1930: 79–80; see also Silberbauer on the G/wi, 1972: 296–297). Thus when Lee and Yellen include a spatial dimension in their consideration of group composition, they discern some persistent association between people and place. It is an association important enough to the !Kung for them to have developed their own concepts for it.

Although Marshall mentioned briefly in her 1960 paper (1960: 325, 330) the two key concepts of owner – *k"xausi* or *kxai k"xausi* (originally spelt *kxau* or *k"ausi*, see Marshall 1960: 325; Lee 1972c: 351) and territory – *n!ore* [or *!nore* or *n!ori* (Wiessner 1977)], it was not until 1976 that she presented an extended discussion (1976: 71, 184–187).[4]

Lee speaks of the owners of a territory as the sibling-cousin group, male and female, who are at the core of each band and are generally acknowledged as the owners of the permanent waterhole focal to it (1976: 77). Although this does not make it entirely clear that everybody, in the past, was an owner of one or other territory, Wiessner and Marshall (1976: 184–185) indicate that this was so with the oldest owner normally being the most important. Ownership automatically confers a joint right to the resources of the territory (Marshall 1976: 189; Wiessner 1977: 49) which others are fearful of trespassing on without permission (Wiessner 1977: 53–54), although large game may be pursued across *n!ore* (Wiessner 1977: 50). However the resources of a *n!ore* are not exclusively reserved for the owners; anyone who has a close relative or exchange partner in camp may, observing elementary good manners, enjoy the resources of the area around the camp[5] (Lee 1976: 78; Wiessner 1977: 50). With marriage a person gains the right to utilise the *n!ore* of the spouse but this lapses if the marriage dissolves (Wiessner 1977: 51). Even travellers and visitors should ask permission before taking water, according to Marshall, because as one !Kung commented, 'the owners know how much there is and how it should be managed' (1976: 190).

More information is available on *n!ore*. All sources agree on the general definition of it as a block of land which contains food resources and water points which together form a basic subsistence area for the resident group (Lee 1976: 77; Wiessner 1977: 49; Marshall 1976: 71–72). Each *n!ore* has a permanent or semi-permanent water source as a dry season base and nearby plant food within range of the water during the dry season. The *n!ore* vary greatly in size and lack clearly defined boundaries, but all habitable parts of the area were part of one or other *n!ore* (Marshall 1976: 71–72; Wiessner 1977). Usually the territory is a single continuous tract of land but it need not necessarily be so. Marshall gives an example in which part of a *n!ore* is in a mongongo nut grove fifty miles distant from the central waterhole, without all the intervening land belonging to the same *n!ore* (1960: 335).

Traditionally everybody had a *n!ore*, although by 1964, in only two cases were groups using their *n!ore* in anything like the traditional manner: Lee writes:

> Even though these semisettled groups spend most of the year at /ai/ai or Tsum!we), each tries to spend at least a month or two in the home n!ore. Unlike the Australians, the !Kung Bushmen do not maintain totemic sites within their home localities. Nevertheless, the ties to the n!ore are certainly based on sentiment as well as economic expediency; this emotional content is expressed in the following quotation from a young woman member of group 3 now living at /ai/ai:'[You see us here today but] you know we are not /ai/ai people. Our true n!ore is East at /dwia and every day at this time of year [November] we all scan the eastern horizon for any sign of cloud or rain. We say, to each other, "Has it hit the n!ore?" "Look, did that miss the n!ore?" And we think of the rich fields of berries spreading as far as the eye can see and the mongongo nuts densely littered on the ground. We think of the meat that will soon be hanging thick from every branch. No, we are not of /ai/ai; /dwia is our earth. We just came here to drink the milk.' (1972b: 142).

Ownership rights to a *n!ore* are usually acquired from either the father or mother. According to both Marshall and Wiessner, a person becomes an owner of the *n!ore* in which the parents are established and with which they are identified (Marshall 1976: 184; Wiessner 1977: 50). Less frequently rights to a *n!ore* may be acquired from a namesake (Wiessner) or by occuption of an empty *n!ore*.

A *n!ore* can only be strongly held, or be a person's 'big *n!ore*', by residence in it, the other parent's *n!ore* will then be a person's 'little *n!ore*' or weakly held (see Marshall 1976: 185; Wiessner 1977: 50). Constant shifting is looked down on — 'What kind of business is that?' one !Kung commented disdainfully to Marshall (1976: 185), — and usually reflects the fact that somebody is difficult to get along with.

Sometimes a *n!ore* becomes vacant because the demographic process has greatly reduced the owners and they have then moved to live with kinsmen elsewhere. Under these conditions people from another area may ask permission of those who hold the *n!ore* weakly for the right to live there. If a man lives for a long period in such a *n!ore* without resident owners he, 'may come to feel like an owner of the *n!ore* where he lives' (Marshall 1976: 185) and his children become owners of it if nobody contests their claim (Wiessner 1977: 51).

Lee provides data on the relationship between the *n!ore* of 151 males over fifteen and those of their parents. Thirty-nine per cent share their *n!ore* with their father, 26.5% with their mother, 10.6% with both partents, 13.9% with neither and 9.3% did not know or have a *n!ore* (see Table I). If Wiessner's information is correct, the 13.9% listed as owning a *n!ore* that is neither the father's nor mother's could be people who had received it from their namesake or succeeded to an empty *n!ore*. As traditionally nobody was without a *n!ore*, those listed as without one are likely to be people who have been absent from the area on wage labour for many years or simply uncooperative interviewees. It is not so clear what to make of 10.6% who claim the same *n!ore* as both parents. They could be the children of marriages within a band, people who shift frequently because of the difficulties they have in getting on with others, or people keeping their options open.

Further extensive but unpublished data on *n!ore* have been collected by Harpending for both sexes, as part of a general physical anthropological survey. His information broadly agrees with Lee's as can be seen from Table I, though there are differences. The correlation between the *n!ore* of fathers and sons in Harpending's data is lower, the percentage whose *n!ore* coincides with both parents

TABLE I

The correlation between the n!ore *of parents and children*
(after Lee 1972b: 129 and Harpending)

Inherited *n!ore* from	Lee Males		Harpending Males		Females	
	N	%	N	%	N	%
Father	60	39.7	67	26.8	79	27.7
Mother	40	26.5	50	20.0	40	14.0
Both F + M	16	10.6	50	20.0	63	22.2
Neither	21	13.9	83	33.2	103	36.1
No *n!ore*	14	9.3	–	–	–	–
Total	151	100.0	250	100.0	285	100.0

is double and the percentage that coincides with neither is three times as great. Most of these differences are probably due to the way the data were collected and analysed. Because no detailed attention had been paid to the concept of *n!ore*, the polysemic and con-textually dependent variations in the meaning of the term do not appear to have been fully understood. Consequently, when asked for their *n!ore*, some people have designated an area and others a particular place within the area. This means that the correlations are lower than they would be had everybody been compared with their parents in terms of a *n!ore* tract rather than some by tract and others by a place within a *n!ore*. Another factor that may have influenced the differences is the inclusion of more young people who now live in centralised villages. Nevertheless, it is clear from both sets of information that the degree of laterality of association runs high: 66.8% for males and 63.9% for females in Harpending's data and 76.8% for males in Lee's. In other words, married couples usually live with one or other of their parents.

Besides collecting information on the correlation of parents' and children's *n!ore*, Harpending also recorded the *n!ore* of birthplace, postmarital-residence and residence at time of interview — see Table II. It can be seen that there is a strong correlation between *n!ore* and birthplace for both sexes with decreasing correlations for postmarital residence and residence at time of interview. These correlations under-line two points. Residence is surprisingly stable since *n!ore* ownership correlates strongly with birthplace,[6] and not all males today move to live with their wife's parents on marriage. Indeed both Marshall (1960: 339) and Wiessner (1977: 312) indicate that approximately

TABLE II

The relationship between an individual's n!ore and the n!ore of birth, postmarital residence and residence at time of interview

	Females %		Males %	
Birth Place	72.6	374 of 515	75.2	353 of 469
Postmarital Residence	46.2	124 of 268	52.4	138 of 263
Interview Residence	27.5	192 of 696	31.9	211 of 661

50% stay put, which the figures in Table 2 tend to confirm. Not much significance is to be attached to the place of residence at time of interview for it does not allow for the great deal of visiting that goes on.

In summary then, there is a pattern of basically ambilateral inheritance of territory in which each !Kung holds one *n!ore* strongly and the other weakly, derived from the place of long term residence of the parents. The *n!ore* are resource nexuses (Silberbauer 1972: 295) capable of supporting a band in most conditions, and the area used by a band for subsistence is coterminous with the area owned by the *k''xausi* among its residents.

Desert Australians

More than a million square miles of Australia is loosely called desert so not unexpectedly there is considerable natural and cultural variation within this area. Rather than nominate a group as more similar either culturally or situationally to the !Kung than the others, I shall discuss territorial organisation with reference to three groups that lie on an ecological and organisational continuum. These are the Pintupi (see Myers nd), the Warlpiri (see Meggitt 1962) and the Aranda (see Strehlow 1965).

The Pintupi inhabit the harshest Australian environment – flat, lacking surface water and dominated by spinifex and sand. The Aranda domain stands in marked contrast. By desert standards it is a rich region with substantial ranges of hills that trap rainfall run-off in sandy river beds, creating long standing pools of surface water that survive all but the worst droughts. The differences are summed up in population densities: Pintupi densities drop as low as one person to 80 square miles (see Long 1971) while Aranda densities may rise as high as one person to 5 square miles (see Strehlow 1965). Between these two extremes lie the Warlpiri with an average density of one person to 35 square miles. In the southern parts of their region, conditions approach those of the Aranda but for the greater part it too is a mass of spinifex and sand without surface water, much like the Pintupi domain.

The pattern of land ownership varies in concert with the environment. A patrilineal grouping is the land owning unit among all three peoples but whereas among the Aranda it is a well developed clan, it is not much more than a weekly developed lineage among the Pintupi. The Arandic clan owns a well defined territory (estate in the Australian terminology — see Stanner 1965) which would appear to usually have been a life space (resource nexus or range),[7] criss-crossed by the tracks of ancestral heroes that set up links between the estates they cross. In the poorer country of the Pintupi and northern Warlpiri, estates tend to focus on heartlands which contain the main permanent waters surrounded by a less well defined tract of land. Although the areas beyond the heartland contain many named places, all of which are owned by one group or another, not enough mapping has been done to make it clear whether these fall into distinct contiguous tracts of land in all cases or whether the sites of different groups intermingle. Further complicating the situation is the Pintupi and northern Warlpiri tendency to give great emphasis to the ancestral tracks which link the heartlands to each other to the neglect of the areal estate interest, where it exists. Nevertheless, it seems that rights arising from tracks are in the final analysis secondary to the patrilineal estate interest.

Ownership confers automatic rights to the estate's resources and an interest in its ritual property. People also have a right to the resources of their mother's clan estate arising from their automatic status as managers of that clan's religious property. Because the managerial relationship is based on biological relationship, it cannot be passed on to descendants unlike the patrilineal right.[8]

All three groups believe that in the remote past their heroic ancestors emerged from the subterranean spirit world, creating the landscape and releasing the life giving forces. These forces are found today at all water sources and are the agents responsible for conception. If the water source was made by a kangaroo ancestor then the life force in the surrounding area will be kangaroo and any child conceived nearby will have a kangaroo conception totem. This links it to the whole line of travel of that ancestor and, more importantly, to the particular location it was conceived at.

Besides the conception totem, each individual has an ancestral clan totem associated with his patrilineal clan estate, no matter where he was conceived. This is the heroic ancestor who founded the clan and created the main sacred places within the estate. It is quite possible for the clan and conception totem to be the same, but even if a person is conceived within his estate it may be that the heroic ancestor that created the particular place within it differs from the clan ancestor. Thus, the conception and clan totem may differ both with respect to ancestor and to location although both are located within the same estate. Many people, however, are conceived on other people's estates in which case the likelihood of the clan and conception totem being the same ancestor is greatly reduced. It can happen that if a person's clan ancestor travelled beyond his estate, an individual may be conceived on his track and thus have the same conception and clan totem although these are associated with different places.

The resource and religious rights conferred by conception in somebody else's clan estate are secondary to those of the owners but capable of conversion to a primary right should the owning clan die out (see Peterson, Keen and Sansom 1977, and Myers nd).

Comparison

Both the !Kung and desert Australians have resource rights to their mother's and father's territory. For a !Kung the rights in one of the parents' territories are stronger than those in the other, but which parent's territory depends on their long-term place of residence. Consequently there is no certain transgenerational continuity of strong rights either patrilineally or matrilineally. For a desert Australian on the other hand, resource rights in the father's territory are always strong and associated with ritual rights both of which are transmitted transgenerationally, while resource rights in the mother's territory are not transmitted in this way.

Although residence has no explicit role in determining rights among desert Australians, there are some interesting congruences between the Australian and !Kung patterns. Among both, men frequently move to live with their wife's parents for a considerable period of time, on first getting married. This is the so-called bride-service among the !Kung (see Marshall 1959: 352; Lee 1972c: 358; Silberbauer 1972: 303 on the G/wi) which appears to be related to the importance of women in the food quest. Among the !Kung this seems to be explicit, for a girl's aging parents may comment to a suitor, as part of the formalities of betrothal, 'You have come to make us poor' (Marshall 1959: 351; see also Schapera 1930: 106 on the Heikum who say something very similar). Presumably, this statement refers to the fact that the girl will now have to feed her husband and may be taken away to another band so that she cannot continue to give food to her parents. However, Marshall (1959: 351) also reports aging parents as saying, 'We are old, we shall soon be old, we need a young man to hunt for us' — emphasising the economic advantages to be gained by having a son-in-law living with them. Although there do not appear to be the cultural formulae for expressing the dependence on daughters and their husbands' labour among Australians, desert women provide the bulk of the diet and Australian sons-in-law, like their !Kung counterparts, are obliged to make regular gifts of meat to their in-laws which they can only do effectively if they are coresident with them. It comes as no surprise, therefore, to find that this pattern of coresidence is one of the most consistent features of bands in Australia (see Peterson 1970; Long 1971).

Thus, a similar techno-environmental situation corresponds to marked similarities in residential association and basic resource rights allocation. The similarity would be increased if it could be shown that desert Australians tend to live either in the husband's or wife's own clan estate. This would then correspond to the !Kung situation where, by the time of marriage, husband and wife will each hold a *n!ore* strongly, one of which will be their place of residence.[9]

In theory, it should be simple to establish a *prima facie* case for where a couple have been living by establishing the estates on which their children are conceived, since a conception totem location must indicate residence in an area, however brief. Unfortunately, little such information is available. Given the argument about postmarital residence patterns, the question resolves itself into where older people live, for that will determine where those Aboriginal men who go to live with their wives are to be found.

There is some general evidence that older men tend to live on their own clan estate which would mean that a man living with his wife's parents could frequently be living in her clan country (e.g. see Myers nd on the Pintupi; Strehlow 1964 on the Aranda; note 10 for the Warlpiri). This is reinforced by the well documented strong emotional tie men have to their own estates which makes them keen to die there (see Peterson 1972: 24 for some references).

Strehlow does provide evidence for conception and ancestral totems for a whole Aranda clan prior to disruption by European settlement, but it is not possible to tell whether those conceived outside their patrilineal estate were conceived on their mother's estate or elsewhere (see 1964: 751–754; 1971: map). From the abbreviated version of the genealogy of the Ellery Creek Honey Ant clan, 13 of the 21 clan members were conceived on the estate and four outside. The information on the other four is insufficient to determine on which estate they were conceived. What is surprising about these figures is the high degree of patrilocal residence they imply, even if the four unknown cases all proved to be outside the clan estate. It also contrasts with the !Kung situation where residence is divided about equally between husband's *n!ore* and wife's *n!ore* (see Marshall 1960: 339; Wiessner 1977: 312). Of course the Aborigines, like the !Kung, move about a great deal, visiting, trading and going to ceremonies. But from a long term perspective the remarkable thing is how persistently the same people are to be found in the same places (for the !Kung see Wiessner 1977: 324–326; for desert Aborigines see note 11).

The other similarity between the !Kung and the Australians is the way they cope with the fluctuations of the demographic process. Because of the small size of land owning groups in each area, their viability can be affected by quite small variations in the sex ratio at birth or in the incidence of fertility and mortality. In consequence, groups frequently die out and others expand and eventually split. The solutions for dealing with this recurrent problem are institutionalised in both Australia and the Kalahari: weak or secondary rights are converted over time into primary rights (see Peterson, Keen and Sansom 1977 for Australia; Marshall 1960: 339; 1976: 196 and Wiessner 1977 for the !Kung; also Silberbauer 1972: 308 for the G/wi).

Ecology and Ideology

Why should such a similarity exist? I suggest that it is because as population densities reach regional limits under the existing technology, land ownership ideologies which exhaustively divide up people and place have adaptive consequences. Specifically, I suggest, they code the distribution of resource nexuses. This not only allows people to predict where other groups are and which resources they are using (see Peterson 1975; Wiessner 1977: 54), thus eliminating direct competition for food, but it also allows for the modified working of the principle of optimum numbers (see

Carr-Saunders 1922) facilitating population regulation.[12]

Although abortion and infanticide are the main ways of controlling numbers, the often-neglected question is how do hunter-gatherers know when to bring these into play and at what levels of intensity in the absence of any widely based collective concern with population density. Emphasis on low fertility combined with variations in work effort that lead individuals to space children widely neglect a crucial point. Hunter-gatherers have supported population expansion: it was they who colonised the world.

The problem is particularly clear with respect to Australia. For several millions of years Australia-New Guinea has been separated from the rest of the world by at least 50 miles of open sea. It seems most unlikely that the 250,000–300,000 Aborigines estimated to have been in Australia at the time of Captain Cook's arrival, were all immigrants rather than the descendants of small handfuls of successful sailors. It might be argued that it had taken the full 40,000 years of known occupation to reach this level and that the population was still growing when Cook arrived, but the evidence makes this seem improbable. Whatever reservations one may have about the details of Birdsell's modelling of the populating of the continent (1957) his basic point seems secure: there must have been, as the archaeology suggests, periods of several thousand years, at the very least, in which the population was in some kind of equilibrium. This is further reinforced by the deterministic relationship Birdsell has shown to exist among desert dwellers and the local rainfall (1953). I suggest that it was the territorial organisation that provided localised feedback making orderly population control possible by creating finite resource areas for definite populations.[13]

If the ideology is coding resource distribution and therefore influencing the number and distribution of bands, this will be indicated if the *average* clan or *n!ore* membership and the *average* band size are similar.

The reason for this is straightforward. If, to take a hypothetical example, a tribe of 500 people is divided into ten clans giving an average of 50 people per clan and if ethnographic reports indicate that the usual size of bands among this tribe is about 50 people, then the number of clans and the number of bands would be the same. The composition of the two could, of course, be expected to vary widely, but given the correlation it would be legitimate to postulate a direct interactional link between the two, particularly if this correspondence is repeated elsewhere.

Meggitt (1962: 32) gives the Warlpiri population at contact as between 1,000 and 1,200 and estimates there were approximately 40 patrilineal clans (1962: 205). This gives an average of 25–30 people to a clan. Data on band size is poor in the Australian desert but 20–25 is a figure widely reported in travellers' accounts (e.g. see Birdsell 1970: 136). Given the approximate nature of the figures, the correlation provides a *prima facie* case for the proposition among the Warlpiri.

No similar calculation can be carried out for the !Kung because there is neither an estimate of the number of *n!ore* nor of the population associated with them prior to the major disruptions of the 1930s. A dubious way of arriving at an estimate of the number of *n!ore* in the interior !Kung area is to divide the total area of the region by the known size of one *n!ore*, fully recognising that *n!ore* vary considerably in size. Yellen gives the area of the Dobe *n!ore* as 320 km² (1977: 54) which divided into the area of the region (see Yellen and Lee 1976: 43) gives 35 *n!ore* or between 30 and 40. Dividing the estimated total population of the area in 1964 by these figures, an average band size of 30–40 is arrived at. This is much larger than the 20.7 (median 19, range 9–52 see Lee 1965: 46) recorded ethnographically by Lee. Marshall provides different figures. She estimates there were 1,000 !Kung in the Nyae Nyae area (this is almost the same as the 'interior area') living in 28 bands, giving an average of 35.7 people per band. It is of interest, however, that the average number in the 14 bands she has most detailed data on was 24 with a range from 8–57, though why this should be so is not clear (1959: 336).

Little can be said of these figures except that they neither confirm nor deny the kind of correlation postulated:[14]

An objection that might be raised against this comparison of average sizes is that it ignores the fact that in the course of the normal demographic process some groups will die out and others expand. This is an important point because it raises issues about the dynamics of cultural process. However, it does not affect the view espoused here, for I have argued that the *number*, not the size of clans or *n!ore* is the regulative feature. As mentioned above, both the !Kung and Australians have mechanisms for dealing with the recurrent problems of the demographic process which allows empty spaces to be refilled with new owners.

It needs to be emphasised that the argument is a normative one, concerned with the modal form of the territorial organisation determined by ecological limits about which fluctuations continually take place. Just as it is recognised that Australian clans or !Kung *n!ore* groups do die out or split, so it is also recognised that some *n!ore* and some Aboriginal estates were amalgamated or divided up at times. But if the underlying logic of the *n!ore* and estates is the resource nexus, defined in terms of least cost subsistence strategies, the modal pattern will keep reasserting itself. Although adequate historical evidence is lacking for the existence of these modal levels, the historical existence of such levels is suggested by the widespread synchronic similarities in band and clan size both within particular regions and more generally by the prevalence of the magic numbers (see Lee and DeVore 1968: 11).

Conclusion

The similarities between !Kung and desert Australian territorial organisation have been obscured in recent writing because the behaviour of the one people has been compared with the ideology of the other. In fact, as a conventional cultural ecological approach would lead one to expect, there are some strong similarities between the two people both in terms of territorial behaviour and ideology. There are, of course, differences as well, the major one being the Australians' elaborate patrilineal religious ideology of land ownership associated with the transmission of resource rights. The reasons for this difference are not immediately obvious but there are some factors which appear to be of significance. If patterns of residence and descent are related, then there may well be greater stability of association between people and place in Australia for an ecological reason.

While many desert Australians live in areas of lower rainfall than the San, their water sources are better distributed. An impervious rock strata lies close to the surface of many parts of the desert, trapping water in soakages 10–20 feet below ground level. This water can be reached at many points by digging wells through the sands which protect it from evaporation. By contrast, the sands of the Kalahari lie on a largely permeable rock formation that allows the water to seep out of reach and makes it available only at central points where the impervious rock comes close to the surface. Thus, major !Kung waterholes are shared by a number of bands in a way not common in the Western Desert. This close proximity of bands to each other, particularly in the dry season, probably means there is more visiting than in the dispersed Australian distribution and more readily available options for permanent residence.

Too much cannot be made of this point, however, for the continuities of overall group composition among !Kung bands sharing a waterhole are striking, as Wiessner's analysis of ten years of /ai/ai censuses show. They demonstrate that the main cause of change is the demographic process itself:

> The observer who left /ai/ai in 1964 and returned in 1974 would find about 40 of the original residents absent due to death or emigration, 80 of the original ones still there as well as approximately 50 new residents who had come in with birth, marriage or immigration (1977: 324).

By swamping ideology with behaviour drained of meaning, the emphasis on flux in hunter-gatherer society is now doing as much to obscure the nature of territorial organisation as the concentration on ideology did in the past. Part of the problem is the overly simple relationship between ideology and behaviour assumed by a cultural materialist research strategy which leads to such odd statements as, 'There is much in Bushman culture about land ownership, land inheritance, and living rights, but these serve to *justify* [emphasis added] . . . underlying ecological processes . . . ' (Yellen and Harpending 1972: 247) or that because it is difficult to predict people's movement they are best conceived as random over an individual's life time (Yellen 1977: 41–43). Movement can only be conceived as random in an environment with a handful of permanent water sources and many highly concentrated food staples by assuming the ecological and cultural constraints that are the focus of interest. By ignoring intention and meaning, important distinctions are collapsed and distance becomes the predominant dimension of spatial behaviour. Only by fully integrating verbal with non-verbal behaviour will the basis be laid for an adequate understanding of particular people's way of life and cross-cultural comparisons become meaning-ful.

Footnotes

This paper had its origins in a seminar I organised on hunter-gatherer territoriality while visiting at the University of New Mexico in 1975. Henry Harpending generously made available his computer tape of !Kung data, and Rocky Chasko extracted and analysed the data on *n!ore*. Then much less was known about *n!ore* than now but when in 1976 Rocky Chasko and Elizabeth Cashdan left for fieldwork in Botswana they managed to glean a little more information, although not working with the !Kung, and feed it back to me through valuable comments on a draft of this paper. Shortly before complet-ing the paper I had the good fortune to run into Polly Wiessner who provided access to her thesis and discussed the !Kung ethnography with me. I am also greatly indebted to Jim O'Connell for his unfailing interest and incisive comments and ideas on Aboriginal territorial organisation. Thanks are due to

Alan Barnard for help with the San literature.

1. Behaviour is here used as a shorthand for non-verbal behaviour when contrasted with ideology.

2. Although Wiessner's thesis goes some way to filling the gap, much remains to be done. In particular, the kind of detailed mapping of the San territories that is now common in Australia would be most useful.

3. See Lee (1965) who suggests that the band may be a statistical artefact of the observer.

4. The term *n!ore* appears in an early dictionary translated as 'country' (see Bleek 1929).

5. The phrasing used by both Lee and Wiessner is unclear. It could be interpreted to imply that owners do not have the right to exclude others from their territory, thereby diminishing the stature of the ownership right. However, simply because people are rarely seen or heard to deny access to others cannot be taken as evidence for the absence of the right to exclude. It not only neglects the fact that those people with close relatives and/or exchange partners in a territory who knew they would be refused access are unlikely to present themselves but also the more subtle metacommunications which might cause visitors to leave. Although Yellen asserts (1977: 40) that Lee has destroyed the notion that there are exclusive territorial rights among the !Kung, this issue requires more subtle and extensive examination than it has received so far. What is clear is that on occasions !Kung do defend their territories (e.g. see Wiessner 1977: 53–54).

6. Since the 1890s Herero have grazed cattle in the Dobe region during the summer. By the middle 1920s they had started to take up permanent residence and today at /ai/ai, for example, there are 600 cattle and 38 Herero (see Wiessner 1977: 19, 66–68; Lee 1976b). Increasingly, the !Kung have come to focus their lives on the permanent waterholes shared with Herero so that for many young people, birthplace and place of residence may coincide more frequently than they did in the past.

7. The relationship between estate and range is one of the issues in the current controversy on Aboriginal territorial organisation. Partly because Radcliffe-Brown assumed they were identical, critics have assumed they were not when the existence of the patrilineal band was denied. A number of authors appear to assume that estates were not big enough to be life spaces, but I think that generally speaking they are wrong. Two points can be made. First, many people assume the range is larger than the estate because they fail to make the distinction between an individual's range and a band's range. Second, while estates are amalgamated and split, I think the modal form is congruent with the range: see pages 123–124.

8. The most readily available account of the managerial relationship is provided by Maddock (1974: 35–36) but it is inaccurate with respect to the Aranda where it is not a voluntary relationship as Maddock, following Strehlow, suggests. This has become quite clear from the recent work carried out by Aranda-speaking missionaries in connection with landrights.

9. The place of residence is the *n!ore* where the couple reside. Since the spouse

living away from his/her *n!ore* cannot be holding it strongly, it means that in one sense one spouse holds one *n!ore* strongly and another weakly while the other holds two weakly. However, by a change of residence, a weakly held *n!ore* can be reconverted to a strongly held one.

10. There is no quantitative evidence on this for the Warlpiri but it is highly significant that with the passing of the *Aboriginal Land Rights (Northern Territory) Act 1976* a number of Warlpiri have been taken back to their own estates to be buried, from the settlement where they now live. This is explicitly so that the owners can point to continued evidence for their right to the land and, more importantly I think, so that the emotional ties are strengthened among the younger people who have always lived on the government settlement and spent little time hunting and gathering in their own country.

11. No published evidence is available but from an analysis of patrol reports into the Western desert during the 1950s and 1960s some will become available shortly.

12. The details of this argument are set out in full in Peterson 1975. The model is decision based and does not assume any broad awareness of a population problem among hunter-gatherers.

13. At first sight this might appear to be no different from Birdsell's own model. It is, however, because his model is based on the patrilineal band encapsulated in its own territory which does not fit with the ethnographic evidence. The argument set out in Peterson 1975 is designed to resolve the difficulty of the non-existence of the patrilineal band in the working of the principle of optimum numbers.

14. Harpending lists 54 *n!ore* but not all fall within the interior !Kung area and some of the names he has as *n!ore* appear only to be places, not tracts of land. It seems likely that *n!ore*, like its Warlpiri counterpart *nguru*, depends greatly on its context for the level of spatial unit being referred to. This may not have been realised.

References

Birdsell, J.B. 1953. Some environmental and cultural factors influencing the structuring of Australian Aboriginal populations. *American Naturalist* 87, 171–207.

——— 1957. Some population problems involving Pleistocene man. *Cold Spring Harbor Symposia on Quantitative Biology* 22, 47–70.

Bleek, D.F. 1929. *Comparative vocabularies of Bushman languages*. Cambridge: Cambridge University Press.

Carr-Saunders, A. 1922. *The population problem*. Oxford: Clarendon Press.

Helm, J. 1969. Remarks on the methodology of band composition analysis. In *Contributions to anthropology: band societies*, National Museums of Canada Bulletin No. 228, Anthropological series No. 84, D. Damas (ed.). Ottawa: National Museums of Canada, 212–217.

Lee, R.B. 1965. *Subsistence ecology of !Kung Bushmen*. Unpublished Ph.D.

thesis, University of California, Berkeley.

Lee, R.B. 1972a. Work effort, group structure and land-use in contemporary hunter-gatherers. In *Man, settlement and urbanism*, P.J. Ucko, R. Tringham and D.W. Dimbleby (eds.). London: Duckworth, 177–185.

—— 1972b. !Kung spatial organization: an ecological and historical perspective. *Human Ecology* 1, 125–147.

—— 1972c. The !Kung Bushmen of Botswana. In *Hunters and gatherers today*, M.G. Bicchieri (ed.). New York: Holt, Rinehart and Winston, 327–368.

—— 1976a. !Kung spatial organisation: an ecological and historical perspective. In *Kalahari hunter-gatherers*, R.B. Lee and I. DeVore (eds.). Cambridge, Mass.: Harvard University Press, 73–97.

—— 1976b. Introduction. In *Kalahari hunter-gatherers*, R.B. Lee and I. DeVore (eds.). Cambridge, Mass.: Harvard University Press, 3–24.

Lee, R.B. and I. DeVore 1968. Problems in the study of hunters and gatherers. In *Man the Hunter*, R.B. Lee and I. DeVore (eds.). Chicago: Aldine, 3–12.

Long, J.P.M. 1971. Arid region Aborigines: the Pintupi. In *Aboriginal man and environment in Australia*, D.J. Mulvaney and J. Golson (eds.). Canberra: Australian National University Press, 262–270.

Maddock, K. 1974. *The Australian Aborigines: a portrait of their society*. Harmondsworth: Penguin Books.

Marshall, L. 1959. Marriage among !Kung Bushmen. *Africa* 29, 335–365.

—— 1960. !Kung Bushman bands. *Africa* 30, 325–355.

—— 1965. The !Kung Bushmen of the Kalahari desert. In *Peoples of Africa*, J.L. Gibbs (ed.). New York: Holt, Rinehart and Winston, 241–278.

—— 1976. *The !Kung of Nyae Nyae*. Cambridge, Mass.: Harvard University Press.

Meggitt, M.J. 1962. *Desert people: a study of the Walbiri Aborigines of central Australia*. Sydney: Angus and Robertson.

Myers, F. nd. *To have and to hold: a study of persistence and change in Pintupi social life*. Unpublished Ph.D. thesis, Bryn Mawr College, Bryn Mawr.

Peterson, N. 1970. The importance of women in determining the composition of residential groups in Aboriginal Australia. In *Woman's role in Aboriginal society*, F. Gale (ed.). Canberra: Australian Institute of Aboriginal Studies.

—— 1972. Totemism yesterday: sentiment and local organisation among the Australian Aborigines. *Man (N.S.)* 7, 12–32.

—— 1975. Hunter-gatherer territoriality: the perspective from Australia. *Am. Anthrop.* 77, 53–68.

Peterson, N., I. Keen and B. Sansom 1977. Succession to land: primary and secondary rights to Aboriginal estates. In *Official Hansard report of the joint select committee on Aboriginal land rights in the Northern Territory*. Canberra: Government Printers. 19th April 1977, 1002–1014.

Radcliffe-Brown, A.R. 1930. The social organisation of Australian tribes, part 1. *Oceania* 1, 34–63.

Schapera, I. 1930. *The Khoisan peoples of South Africa: Bushmen and Hottentots*. London: George Routledge and Sons.

Service, E.R. 1965. *Primitive social organization: an evolutionary perspective.* New York: Random House.

Silberbauer, G. 1963. Marriage and the girl's puberty ceremony of the G/wi Bushmen. *Africa* 33, 12–24.

—— 1972. The G/wi Bushmen. In *Hunters and gatherers today,* M.G. Bicchieri (ed.). New York: Holt, Rinehart and Winston, 271–326.

Stanner, W.E.H. 1965. Aboriginal territorial organisation: estate, range, domain and regime. *Oceania* 36, 1–26.

Steward, J.H. 1936. The economic and social basis of primitive bands. In *Essays in honor of A.L. Kroeber.* Berkeley: University of California Press, 331–350.

Strehlow, T.G.H. 1964. Personal monototemism in a polytotemic community. In *Festschrift für Ad. Jensen, 2 vols.,* E. Haberland, M. Schuster and H. Straube (eds.). Munich: Klaus Renner, 723–754.

—— 1965. Culture, social structure and environment in Aboriginal central Australia. In *Aboriginal man in Australia,* R.M. and C.H. Berndt (eds.). Sydney: Angus and Robertson, 121–145.

—— 1971. *Songs of central Australia.* Sydney: Angus and Robertson.

Turnbull, C. 1968. The importance of flux in two hunting societies. In *Man the hunter,* R.B. Lee and I. DeVore (eds.). Chicago: Aldine, 132–137.

Wiessner, P. 1977. *Ilxaro: a regional system of reciprocity for reducing risk among the !Kung San.* Unpublished Ph.D. thesis, University of Michigan.

Williams, B.J. 1968. The Birhor of India and some comments on band organization. In *Man the hunter,* R.B. Lee and I. DeVore (eds.). Chicago: Aldine, 126–131.

—— 1974. *A model of band society.* Memoir 29 of the Society for American Archaeology.

Woodburn, J. 1968. Stability and flexibility in Hadza residential groupings. In *Man the hunter,* R.B. Lee and I. DeVore (eds.). Chicago: Aldine, 103–110.

Yellen, J.E. nd. Post-pleistocene hunter gatherer adaptations to desert environments. *Paper presented to the Smithsonian Conference on Human Biogeography,* April 1974, 1–32.

—— 1977. *Archaeological approaches to the present: models for reconstructing the past.* New York: Academic Press.

Yellen, J.E. and H. Harpending 1972. Hunter-gatherer populations and archaeological inference. *World Archaeology* 4, 244–253.

Yellen, J.E. and R.B. Lee 1976. The Dobe-/Du/da environment: background to a hunting and gathering way of life. In *Kalahari hunter-gatherers,* R.B. Lee and I. DeVore (eds.). Cambridge, Mass.: Harvard University Press, 27–46.

KALAHARI BUSHMAN SETTLEMENT PATTERNS[1]

ALAN BARNARD

Introduction

In recent years, a great deal has been written about Kalahari Bushman (or San) ecology. Nevertheless there has been virtually no attempt at comparison of the different settlement patterns found among Kalahari Bushman peoples. Instead, environmental determinists have merely assumed, without making any comparisons, that environment nearly always determines the most important elements of social organisation (e.g. Lee 1965; Silberbauer 1973). Or when inclined to make comparisons, scholars have tended to compare only a single Bushman society to hunting and gathering societies in other parts of the world (e.g. Bicchieri 1969; cf. Nicolas Peterson's paper in this volume). Such comparisons fail to take into account the significant differences in local environment within the Kalahari. For example, John Yellen (1976: 48), writing on Zhu/twasi settlement patterns, has noted that some of the Aborigines of the Western Desert of Australia have the reverse seasonal cycle of the Zhu/twasi !Kung (cf. Gould 1969). 'To understand why such different patterns exist,' Yellen writes, 'it is necessary to examine the Kalahari and Western Australian Desert environments, the distribution of water, plant, and animal resources, and consequently the differing subsistence strategies employed by the !Kung and the aborigines (sic)'. Had Yellen been following the approach I am advocating here, he would have reported that this Western Desert seasonal cycle also exists in the Kalahari, less than 400 kms from his own Zhu/twasi fieldwork area. For like the Western Desert Aborigines mentioned by Yellen, the G/wikhwe and G//anakhwe Bushmen aggregate in the wet season and disperse in the dry season. And incidentally, some desert-dwelling Australian Aboriginal groups do have much the same seasonal cycle as the !Kung, Western Desert wet season aggregations being of short duration and primarily for ritual purposes. (Nicolas Peterson, personal communication.) The reasons for such differences cannot be found by comparing a single Bushman society to a single Aboriginal society, but rather by comparisons within geographical or culture areas.

This paper is comparative but geographically limited. My approach here is intended as an illustration of the kind of study which I think can best test the limits of environmental determinism. I shall describe the settlement patterns of four Kalahari Bushman peoples and note the micro-environmental differences which seem to influence them. My theoretical inspiration is the work of the pioneer of southern African social ecology, Agnes Winifred (Tucker) Hoernle (1923), who made similar comparisons among Nama Khoikhoi groups in Namibia in the early 1920s.

The Kalahari Bushmen

Bushman groups aggregate and disperse according to custom and social circumstance, the availability of plant and animal foods and most importantly, the availability of water. Each ethnic group has its own yearly cycle and its own definition of the units (or levels) of social and territorial organisation.

This paper deals with four peoples (1) the Zhu/twasi !Kung of the northern Kalahari; (2) the east central Kalahari groups (who include the closely related G/wikhwe and G//anakhwe); (3) the !Ko or western ≠ Hoa of the south-central Kalahari, and (4) the Nharo (Naron, //Aikwe) of the central western Kalahari (see map, Figure 1). All of these peoples are believed to be longstanding inhabitants of the Kalahari desert. The only region of the desert that will not be treated in this paper is the far southern region, which is very dry and barren, largely uninhabited, and about which little of relevance has been written.

The Zhu/twasi

The Zhu/twasi, sometimes called by the more generic term '!Kung', are perhaps the best-known of all Bushman peoples. In order to make it clear that I am speaking here only of the central !Kung, and not of the !O !Kung of Angola or the (southern) ≠Au//eisi !Kung, I shall refer to them as Zhu/twasi, the term by which the central !Kung prefer to call themselves. The Zhu/twasi number some 7,000 people (Lee 1965: 13), most of whom live in close contact with Herero and Tswana cattle herders. Even so, hunting and gathering is prevalent in most Zhu/twasi areas and their traditional lifestyle has been well-documented. The following description is a composite, based on the work of Lorna Marshall (1976), Richard B. Lee (1965; 1968; 1969;

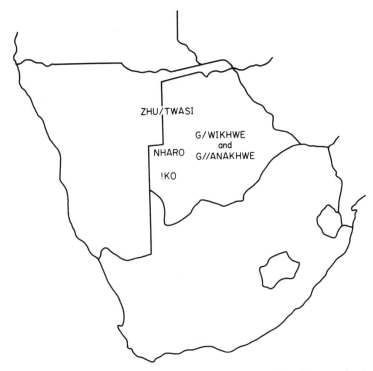

Fig. 1. Map of Southern Africa, showing approximate locations of the four peoples described.

1972a; 1972b; 1972c; 1972d; 1973; 1974), John Yellen (Yellen and
Lee 1976; Yellen 1976; 1977), Patricia Draper (1975) and Polly
Wiessner (1977).

Zhu/twasi country has a greater abundance of resources than any
other part of the Kalahari. The land is covered with acacia trees and
a variety of edible root plants, and even supports large shade trees
which grow up to fifteen metres in height. In the Dobe area, nearly
500 different species of plants and animals are known and named
by the inhabitants. Of these, some 150 plant species and 100 animal
species provide food, clothing and utensils to serve virtually all their
needs (Lee 1972a: 342; Yellen and Lee 1976: 37–38). Vegetable
foods include the famous mongongo or mangetti nut (!Kung name:
//"xa; scientific name: *Ricinodendron rautanenii* Schinz), as well as
other nuts, roots, berries, fruits and leaves. The mongongo is particularly nutritious and abundant; Lee (1973: 313) found at least
one mongongo grove within ten kilometres of each major waterhole
in the Zhu/twasi area he studied. The nuts can be collected throughout

the year and are especially sought during the relatively lean, dry months from April to September.

The availability of these foods, and water resources, is an important determining factor of Zhu/twasi residential and social organisation, and the Zhu/twasi have a loose socio-territorial structure which is ideally suited to their subsistence needs. There are only two distinct units of socio-territorial organisation, the nuclear or extended family and the band. Each family occupies a hut or space without shelter (e.g. beside a tree) in any campsite or settlement. The groups of families who camp and travel together in search of food and water form the band (or *n//abesi*). Bands are fluid in membership, but the resource bases on which they depend – waterholes and mongongo groves – are permanent (Marshall 1976: 187–91). Although Yellen (1977: 43) has put forward the idea of an 'almost random' movement of individuals around Zhu/twasi country, other ethnographers and the Zhu/twasi themselves (Yellen 1977: 41) prefer to represent socio-territorial organisation in terms of a band model.

Each band occupies a recognised territory (or *n!ore*) of some 300 to 1,000 km² (approximated from map, Yellen 1976: 55; 1977: 39). The territories each centre around one or more waterholes and often slightly overlap those of other bands. Thus areas of good resources may be used by more than one band. Other areas may not be used at all, and their marginal resources may remain unclaimed by any band (Lee 1965: 147–48; Yellen 1976: 54–59; 1977: 37–49). As one Zhu/twasi man told Lorna Marshall: 'What good is ground that produces no food? One cannot eat it' (Marshall 1976: 72).

For a few months during the dry season, often two or more bands will share the same waterhole; and in the wet season, each band goes its own way in order to exploit the rainwater which collects in tree trunks and pans throughout the country. Today, due partly to outside pressures (including the enlargement of waterholes and the introduction of livestock by Bantu-speaking herders), some bands remain at the permanent waterholes for most of the year (see e.g. Lee 1972b: 183–84; 1972c: 140–42).

In each band there is one individual (the *k"xau n!a*) who is said to be the hereditary 'owner' of all the plants and water (but not the large game animals) which are found in the territory. But in practice, of course, the group as a whole controls the resources and the *k"xau n!a* has little more power or influence than any other band member (see Marshall 1976: 191–95).

Children are named after their grandparents or other close relatives, and in relation to any given ego, all Zhu/twasi are classified as members of egocentric kin categories by rules of namesake-equivalence (see Marshall 1976: 223–42). Yet Zhu/twasi bands are made up of consanguineously and affinally related kinsmen. Although bride service is practised, band members are recruited bilaterally (through either males or females), and each band normally has at its core a senior group of siblings from whom most band members are descended (see Lee 1972a: 350–56; 1974: 169; Marshall 1976: 182–84; Wiessner 1977: 48–59). In 1952 the nineteen bands of the Nyae Nyae (N//hwa!ai) area averaged about twenty-five people each; the largest distinct band included forty-two people and the smallest included only eight (Marshall 1976: 196). Bands do not seem to vary in size according to the season, although long-term visiting and changes in band membership are frequent. Interaction between members of different bands tends to be on an individual and not on a band basis (Marshall 1976: 199–200; Yellen 1977: 41–47).

The G/wikhwe and G//anakhwe

The G/wikhwe and G//anakhwe, sometimes together called the 'Central Kalahari Bushmen', are almost indistinguishable from each other in language, culture and settlement pattern. The G//anakhwe (or G//ana) inhabit mainly the eastern area of Botswana's Central Kalahari Game Reserve, and the G/wikhwe (or G/wi) inhabit mainly the southwestern area of the Reserve. But for our purposes here they will be considered together as one people. Two major studies of their ecology have been carried out, one on the G/wikhwe by the former Ghanzi district commissioner and Bushman Survey Officer George B. Silberbauer (1965; 1972; 1973), and the other on the G//anakhwe by the Japanese anthropologist and primatologist Jiro Tanaka (1969; 1976; m.s.). Here I shall rely primarily on Silberbauer's work, since he has concentrated more than Tanaka on settlement patterns.

The Bushman inhabitants of the Central Kalahari Game Reserve are less numerous than the Zhu/twasi – recent estimates range from 1,000 (Tanaka 1976: 100) to 3,500 (Guenther 1974: 41). They speak a language of the Khoi or 'Hottentot' family and thus are linguistically unrelated to the !Kung. Nevertheless, they have a great deal in common with the !Kung groups with regard to technology, religion and other aspects of culture. And of the four peoples discussed in this paper, they have been the least influenced by the Bantu-speaking herders.

The Game Reserve extends over more than 50,000 km² and consists of three environmental zones: (1) an area of sand dunes covered with a variety of trees and shrubs in the north, (2) scrub woodlands in the south, and (3) a central scrub plain. Of these three zones, the plain and the southern woodlands are exploited by the Bushmen. The northern dune woodland, although richest in large game and in seasonal water resources, does not have enough edible plants to support a permanent human population. Each band occupies a territory extending some ten to fifteen kilometres from its centre. The six G/wikhwe band territories surveyed by Silberbauer (1972: 295) range from about 450 to 1,000 kms², with an average of just over 750 kms² per band. The bands range in size from twenty-one to eighty-five people with an average size of fifty-seven.

Silberbauer (1972: 296–97) records four patterns of territorial occupation: (1) movement of an entire band within its own territory (syneocious migration), (2) temporary migration of the band to an adjacent band territory (synoecious emigration), (3) segmentation forced upon the band in times of severe drought, and (4) seasonal segmentation or temporary dispersal (cf. Tanaka 1969: 12–14). From the time of the first rains (usually about November) until mid-winter (July), each synoecious band migrates as a unit within its defined territory. After the rain water has disappeared, the synoecious band exploits the water-bearing tsama melon crop. Tsama melons are usually found in several places within any band territory (although their abundance varies from year to year), and the band migrates from one tsama patch to the next. In the very dry winter months, beginning in July or August, the bands split into family units and spread throughout each band territory. Each family exploits a different part of the territory and remains there until the lean months are over. Succulent plants and liquids in the bodies of animals hunted are their only sources of water for several months. Then some time after the first rains (about December), the bands re-form to enjoy again the season of relative abundance.

The reason that the Zhu/twasi maintain high-density population during the dry season is that ample water and ample food resources, particularly mongongo nuts, are obtainable only at specific locations. But unlike the Zhu/twasi, the G/wikhwe and G//anakhwe have no permanent waterholes and no staple food supply. Mongongo trees are not found anywhere in the Central Kalahari Game Reserve, or indeed anywhere at all except in some parts of Zhu/twasi country. For this reason, the G/wikhwe and G//anakhwe must rely on a wide variety — some seventy-nine species — of seasonal plant foods. Most of these are obtainable for only a few months of each year (Tanaka 1976: 105–09, 117–18; Silberbauer 1972: 282). Most plant foods, tsama melons excepted (they grow in patches), are scattered over the countryside, and G/wikhwe and G//anakhwe seasonal migratory dispersals are directly related to the exploitation of the scarce resources of their environment (see maps, Silberbauer 1972: 296–97; or 1973: 280–81).

As among the Zhu/twasi, recruitment to the band is either by privilege of ownership of the territory (inherited through one's father or mother) or through in-marriage. Most marriages are band-exogamous and uxorilocal residence is the general rule, at least until the birth of a couple's first child. After that, a couple will normally return to the husband's band (Silberbauer 1972: 303). Although they lack the Zhu/twasi naming system, virtually all G/wikhwe and G//anakhwe who come in contact are kin (or are called by kin terms), and bands are allied by ties of marriage and descent to others nearby. Non-allied bands are not enemies but merely kin whose relationship is more distant or more strained. Allied bands help each other in times of need and their members, either individually or collectively, make visits in order to exchange news, arrange marriages and offer mutual support (Silberbauer 1972: 302–04).

The !Ko

The !Ko live in the vast and relatively barren south-central Kalahari, southwest of the Central Kalahari Game Reserve. They are few in number but speak a variety of dialects and are thinly scattered over a territory of nearly 250,000 kms^2; their dialects are unrelated to those of the surrounding Bushman and non-Bushman peoples (see Traill 1974: 8–24). Less data are available on the !Ko than on the Zhu/twasi or the Central Kalahari Bushmen. The main source on their settlement patterns is the work of the German parasitologist and ethologically-oriented anthropologist H.J. Heinz (especially 1966; 1972).

!Ko country is poorer than that of the other major Bushman peoples. There is no staple food plant, and the availability of edible plants and animals, and of water, fluctuates considerably. Traditionally, before the recent establishment of Botswana government boreholes, the !Ko had to rely on seasonal rainwater which collects in the large flat pans and valleys found in the region, and like the Central Kalahari Bushmen, on tsama melons and other water-bearing plants (Heinz and Maguire n.d.: 39–40).

The !Ko are highly territorial, they have a well-known dislike of strangers (see Heinz 1975), and seasonal changes do not seem to have much effect on their social organisation. According to Heinz (1972: 406–10), each band disperses 'periodically' and each family exploits

its own area within the band territory, although by right, all band territory is accessible to the entire band. In addition, adjacent !Ko bands are grouped into a larger territorial unit which Heinz has called the 'nexus' and which I term the 'band cluster'. Thus among the !Ko there are three levels of territorial organisation (the family, the band and the band cluster), as opposed to only one (the band) among most other Bushman peoples today. The existence of a nuclear family level of territorial organisation must cause the !Ko some problems in the transmission of primary rights to territory from one generation to the next, but not enough data are available to say exactly how this happens.

Among the !Ko, close kinship ties unite bands of the same band cluster and great animosity may exist across band cluster boundaries. Recruitment to band membership is by birth or in-marriage and by permission of the headman. Bride service and uxorilocal residence are practised, and band clusters tend to be endogamous. When, on rare occasions, marriages are contracted with individuals outside the band cluster, membership does not change. Each partner in theory remains a member of his or her natal band cluster; the in-marrying partner is a foreigner in his spouse's country (Heinz 1966: 108). Band cluster boundaries are often dialect boundaries, and in stark contrast to the Zhu/twasi, their territories are bounded by a strip of 'no man's land' between them (compare Fourie 1928: 85). On hunting and gathering expeditions this strip is generally avoided. Only plant resources and not animals are owned by the band, and with permission, members of one band may hunt in the territory of another band of the same cluster (Heinz 1966: 91ff; 1972: 408–10). The only apparent seasonal aggregation for the !Ko is a ritual rather than an environmental one: in the autumn, before the harsh dry season, each band cluster aggregates for its boys' initiation ceremony (Heinz 1966: 125–34).

The Nharo

The 5,000 Nharo live in a comparatively well-watered region of the Kalahari, along the limestone ridge which stretches roughly from south of the Lake Ngami in the northeast, to west of the Botswana/Namibia border at Mamono in the southwest. The Nharo were studied briefly in the 1890s by Siegfried Passarge (1907 *passim*), in the early 1920s by Dorothea Bleek (1928), and in the 1970s by me.[2] Unless otherwise stated, the material presented here is my own (cf. Barnard 1976: 88–98, 234–37).

In most areas a great variety of seasonal plant foods is available, and in the south various kinds of antelope and smaller animals are hunted, although the great herds of elephant and other big game reported by the nineteenth century gentleman adventurers (Andersson 1856; Baines 1864; Chapman 1868; Schinz 1891) have now disappeared.

Natural water was plentiful on Ghanzi ridge during the late nineteenth century (see Hahn 1895: 616), and according to several of my informants, the Nharo used to spend the dry season camped at the large permanent waterholes and the wet season scattered among the many seasonally-filled pans. In other words, their seasonal cycle resembled that of the Zhu/twasi, rather than that of the Central Kalahari Bushmen, to whom the Nharo are more closely related, linguistically and culturally. Like the !Ko, they are divided into band clusters. Each band (*tsou*) and each band cluster (*n!u*) is named, but cluster boundaries do not distinguish separate dialects and, perhaps more today than in the past, are not always clearly defined geographically. Unlike the !Ko, the Nharo are known for their hospitality. The Kalahari Afrikaners still tell of the time (in the late 1890s) when the Nharo of Ghanzi fed and looked after a small band of Afrikaner women and children for two years, while the Afrikaner menfolk scouted the territory father north.

Dorothea Bleek (1928: 4), who studied the western Nharo in the early 1920s, described them as living in small settlements, each within a short distance of a waterhole. This distance was always less than an hour's walk. Settlements were not located at the waterhole itself because of the possibility of frightening away the game which would come to drink there. Sometimes several bands, or settlements, would share the same waterhole; and these settlements would include anywhere from three to twenty huts, each occupied by two to five people.

The present situation is different in many areas because of the extreme over-abundance of water. Although the water table has sunk in the past few decades (particularly in the 1950s and 1960s), Afrikaner ranchers, Kgalagari subsistence herders, the Botswana government and the Ghanzi District Council have dug many wells and boreholes. Today these, and not the permanent natural waterholes, supply most of the water used by the Nharo. In the northeastern part of the ranching area, where boreholes are most numerous, Nharo have settled in small nuclear or extended family units, with one or two families at each borehole. In the southwest, bands are larger. The average band size in the southern band cluster territory where I worked (N//wa //xe) is about twenty. There, bands range in size from eight to over forty people, and in a few cases two or three bands share the same waterhole or borehole. Their territories are considerably smaller than those of other Bushman groups, sometimes only about thirty square kilometres (or half the size of the average ranch) in the interior of the ranching area. In the southern area they are often much larger, but not always precisely bounded. Where settlements are close together, sharing the same waterhole, territories may be utilised collectively by more than one band.

Each band has a headman (or //*eixaba*), who by definition is simply the eldest male resident of the settlement. He is not necessarily a member of its genealogical core. As among the Zhu/twasi, the headman is said to be the 'owner' of the flora but not the fauna of 'his' territory. He has little (if any) political power, and quite often other men, such as particularly good hunters or well-known medicine men, command greater respect in the community.

Visiting is frequent. Some Nharo are compulsive visitors, leaving their homes every three or four days in order to see their friends and relatives elsewhere. Hardly a month goes by in which band composition does not change. Yet band location changes much less frequently. Settlements are found near each waterhole and borehole and in general can be occupied all year round. Band membership may change, old huts may be knocked down and new ones put in their place, but G/wikhwe-style synoecious migrations are (and were in the past) unnecessary.

Membership of a band is open to anyone of the same cluster or anyone who has married into it, although as among the !Ko, most marriages do take place within the cluster. The Nharo classify all members of their society as kin through rules of namesake-equivalence, a custom probably borrowed from the !Kung (see Barnard 1976: 158–66). Yet it may be significant to note that the same set of personal names is found throughout Nharo country, whereas !Kung names are more localised. This may reflect the greater geographical mobility achieved by the Nharo, which in turn is partly due to the abundance of water along Ghanzi ridge.

Summary and Conclusion

The four peoples described in this paper appear to have very different settlement patterns. Some of these apparent differences may partly reflect the diverse theoretical perspectives of the various ethnographers (e.g. Heinz's concern with territoriality), but even so, a few general conclusions can be drawn.

The Zhu/twasi aggregate in the dry season and disperse in the wet

season, while the Central Kalahari Bushmen disperse in the dry
season and aggregate in the wet season. As Megan Biesele (1971: 65)
has noted, water (and not the presence of game or plant resources) is
the primary reason for seasonal migrations among both the Zhu/twasi
and the G/wikhwe (cf. Yellen 1977: 64). !Ko and Nharo settlement
patterns show less seasonal variation, yet the !Ko appear to be
highly territorial and the Nharo (except perhaps in the drier
western areas), relatively unconcerned with territoriality. The !Ko
have fixed territorial boundaires, but the Nharo have flexible
boundaries and fixed settlements. They can remain at their waterholes
all year round, or like the Zhu/twasi, they can exploit outlying
seasonal water supplies after the rains.

Both the !Ko and the Nharo recognise a level of territorial
organisation larger than the band, but the significance of this level –
the band cluster – is quite different for the two peoples. !Ko rarely
travel beyond their band cluster boundaries, but many Nharo do,
and they frequently extend their social networks to individuals in
other band clusters. In Yellen and Harpending's (1972) terms, !Ko
bands and band clusters are highly 'nucleated' and Nharo bands
and band clusters (like Zhu/twasi and, to a lesser degree, G/wikhwe
and G//anakhwe bands) tend towards the 'anucleate' end of the
spectrum. From most nucleated to least nucleated, the four groups
can be ranked: !Ko, Central Kalahari Bushmen, Zhu/twasi, Nharo.
From least water to most water, the ranking is the same.

To put it simply (and crudely), there is a direct correlation
between the availability of water and the fluidity of Kalahari Bushman
socio-territorial organisation. Different local environments are
associated with different settlement patterns, and the type and
frequency of water (or water-bearing) resources appears to have a
considerable effect on Bushman dispersals, aggregations and
migrations, notions of territoriality, and perhaps even visiting
customs. But the relationship may be quite subtle. Trying to decide
what is 'purely' cultural and what is dictated by the natural environment is probably a futile task, although the influence of the environment can be seen when the right comparisons are made.

Comparisons are essential if we are to establish ecological correlations. My belief is that the most revealing comparisons are
generally those which can be made within a culture area rather than
between culture areas, and in this paper I have approached the
subject of settlement patterns from that angle. Perhaps elsewhere
than in the Kalahari (or southern Africa generally), water is more
plentiful and therefore less important as a determinant of social

organisation. But comparison of, for example, just one Pygmy or
Eskimo society to one Bushman group will not show this. Scholars
interested in ecological theory or in archaeological analogy would do
well to remember that the Zhu/twasi !Kung are not the only hunter-
gatherers in the world. They are not even the only Bushmen in the
Kalahari.

Footnotes

1. I would like to thank Nicolas Peterson and Polly Wiessner for their helpful
   comments on an earlier draft.
2. My fieldwork among the Nharo was carried out between May 1974 and
   September 1975. I am grateful to the Swan Fund for financial support and
   to the Office of the President of Botswana for permission to conduct this
   work.

References

Andersson, C.J. 1856. *Lake Ngami; or explorations and discoveries during four
   years' wanderings in the wilds of south western Africa.* London: Hurst and
   Blackett.
Baines, T. 1864. *Explorations in south-west Africa.* London: Longman, Green,
   Longman, Roberts and Green.
Barnard, A.J. 1976. *Nharo Bushman kinship and the transformation of Khoi
   kin categories.* Unpublished Ph.D. thesis, University of London.
Bicchieri, M.G. 1969. A cultural ecological comparative study of three African
   foraging societies. In *Contributions to anthropology: band societies*
   (National Museums of Canada Bulletin No. **228**), D. Damas (ed.). Ottawa:
   National Museums of Canada, 172–79.
Biesele, M. 1971. Hunting in semi-arid areas – the Kalahari Bushmen today. In
   Proceedings of the Conference on Sustained Production in Semi-Arid Areas
   (Botswana Notes and Records Special Edition No. 1). Gaborone: Botswana
   Society, 62–67.
Bleek, D.F. 1928. *The Naron: a Bushman tribe of the central Kalahari.* Cambridge:
   University Press.
Chapman, J. 1868. *Travels in the interior of South Africa.* 2 vols. London: Bell
   and Daldy.
Draper, P. 1975. !Kung women: contrasts in sexual egalitarianism in foraging
   and sedentary contexts. In *Toward an anthropology of women,* R.P. Reiter
   (ed.). New York and London: Monthly Review Press, 77–109.
Fourie, L. 1928. The Bushmen of South West Africa. In *The native tribes of
   South West Africa,* by C.H.L. Hahn, V. Vedder and L. Fourie. Cape Town:
   Cape Times, 79–105.
Gould, R.A. 1969. Subsistence behaviour among the Western Desert Aborigines
   of Australia. *Oceania* **39**, 253–74.
Guenther, M.G. 1974. *Farm Bushmen: socio-cultural change and incorporation
   of the San of the Ghanzi District, Republic of Botswana.* (Unpublished report

to the Botswana government.)

Hahn, J.T. 1895. Who is the lawful owner of Ghanzi? (A letter from Hahn to the Imperial Secretary, Government House, Cape Town.) Colonial Archives (London) file no. CO 417–142 (CO 16669), 611–18.

Heinz, H.J. 1966. *The social organisation of the !Ko Bushmen.* Unpublished M.A. thesis, University of South Africa, Pretoria.

—— 1972. Territoriality among the Bushmen in general and the !Ko in particular. *Anthropos* 67, 404–16.

—— 1975. Acculturation problems arising in a Bushman development scheme. *South African J. of Sci.* 71, 78–85.

Heinz, H.J. and B. Maguire n.d. *The ethno-biology of the !Ko Bushmen: their ethno-botanical knowledge and plant lore* (Botswana Society Occasional Paper No. 1). Gaborone: Botswana Society.

Hoernle, A.W. 1923. South-West Africa as a primitive culture area. *South African Geographical Journal* 6, 14–28.

Lee, R.B. 1965. *Subsistence ecology of !Kung Bushmen.* Unpublished Ph.D. thesis, University of California at Berkeley.

—— 1968. What hunters do for a living, or, how to make-out on scarce resources. In *Man the hunter*, R.B. Lee and I. DeVore (eds.). Chicago: Aldine, 30–48.

—— 1969. !Kung Bushman subsistence: an input-output analysis. In *Contributions to anthropology: ecological essays* (National Museums of Canada Bulletin No. 230), D. Damas (ed.). Ottawa: National Museums of Canada, 73–94. (Also reprinted in *Environment and cultural behavior*, A.P. Vayda [ed.]. New York: Natural History Press, 47–79.)

—— 1972a. The !Kung Bushmen of Botswana. In *Hunters and gatherers today*, M.G. Bicchieri (ed.). New York: Holt, Rinehart and Winston, 327–68.

—— 1972b. Work effort, group structure and land use in contemporary hunter-gatherers. In *Man, settlement and urbanism*, P.J. Ucko, R. Tringham and D.W. Dimbleby (eds.). London: Duckworth, 177–85.

—— 1972c. !Kung spatial organization: an ecological and historical perspective. *Human Ecology* 1, 125–47. (Also reprinted in *Kalahari hunter gatherers: studies of the !Kung San and their neighbors*, R.B. Lee and I. DeVore [eds.]. Cambridge, Mass. and London: Harvard University Press, 73–97.)

—— 1972d. Population growth and the beginnings of sedentary life among !Kung Bushmen. In *Population growth: anthropological implications*, B. Spooner (ed.). Cambridge, Mass.: MIT Press, 329–42.

—— 1973. Mongongo: the ethnography of a major wild food resource. *Ecology of Food and Nutrition* 2, 307–21.

—— 1974. Male-female residence arrangements and political power in human hunter-gatherers. *Archives of Sexual Behavior* 3, 167–73.

Marshall, L. 1976. *The !Kung of Nyae Nyae.* (Including reprints of earlier papers published in *Africa*.) Cambridge, Mass. and London: Harvard University Press.

Passarge, S. 1907. *Die Buschmanner der Kalahari.* Berlin: Dietrich Reimer.

Schinz, H. 1891. *Deutsch-Sudwest-Afrika.* Oldenburg and Leipzig: Schultzesche.

Silberbauer, G.G. 1965. *Report to the government of Bechuanaland on the*

144   *Alan Barnard*

Bushman survey. Gaberones: Government Printer.
Silberbauer, G.G. 1972. The G/wi Bushmen. In *Hunters and gatherers today*, M.G. Bicchieri (ed.). New York: Holt, Rinehart and Winston, 271–326.
—— 1973. *Socio-ecology of the G/wi Bushmen.* Unpublished Ph.D. thesis, Monash University, Clayton, Victoria, Australia.
Tanaka, J. 1969. The ecology and social structure of the Central Kalahari Bushmen: a preliminary report. *Kyoto University African Studies* 3, 1–26.
—— 1976. Subsistence ecology of the Central Kalahari San. In *Kalahari hunter-gatherers: studies of the !Kung San and their neighbors*, R.B. Lee and I. DeVore (eds.). Cambridge, Mass. and London: Harvard University Press, 98–119.
—— m.s. *The Bushmen.* (Translated from the 1971 Japanese edition by D.W. Hughes and G.L. Barnes.)
Traill, A. 1974. *The complete guide to the Koon.* Johannesburg: African Studies Institute (University of the Witwatersrand).
Wiessner, P.W. 1977. *Hxaro: a regional system of reciprocity for reducing risk among the !Kung San.* 2 vols. Unpublished Ph.D. thesis, University of Michigan, Ann Arbor.
Yellen, J.E. 1976. Settlement patterns of the !Kung: an archaeological perspective. In *Kalahari hunter-gatherers: studies of the !Kung San and their neighbors*, R.B. Lee and I. DeVore (eds.). Cambridge, Mass. and London: Harvard University Press, 47–72.
—— 1977. *Archaeological approaches to the present: models for reconstructing the past.* New York, San Francisco and London: Academic Press.
Yellen, J.E. and H.C. Harpending 1972. Hunter-gatherer populations and archaeological inference. *World Archaeology* 4, 244–53.
Yellen, J.E. and R.B. Lee 1976. The Dobe-/Du/da environment. In *Kalahari hunter-gatherers: studies of the !Kung San and their neighbors*, R.B. Lee and I. DeVore (eds.). Cambridge, Mass. and London: Harvard University Press, 27–46.

ECOLOGICAL VARIATION ON THE NORTHWEST COAST: MODELS FOR THE GENERATION OF COGNATIC AND MATRILINEAL DESCENT

DAVID RICHES

Introduction

Anthropologists working on nomadic hunter-gatherers are accustomed to the view that the fundamentally 'precarious' circumstances of the hunger-gatherer productive economy inform highly 'flexible' social structures; in hunter-gatherer social structure, flexibility refers basically to the reckoning of kinship obligations bilaterally, the lack of descent group corporations, and the absence of notions of ownership over territory and wild resources (Peterson, this volume). Nowhere has this view been more strongly expressed than in studies of arctic and subarctic hunter-gatherers; in high northern latitudes, economic uncertainty is held to stem largely from the instability and/or fragility in the ecosystem (Dunbar 1968; 1973; cf. Helm 1965; Riddington 1969; Guemple 1972).

Whether or not these arguments hold merit (and I believe that there is evidence that they may not do so entirely[1]), they certainly point to a striking contrast between the social structures of the nomadic hunter-gatherer and the social structures of the 'sedentary' hunter-gatherer on the American Northwest coast. With respect to 'resource groups' (i.e. groups of people concerned with the exploitation of resources within culturally recognised territorial areas), Northwest coast social structures evidence an increasing *'rigidity'* as one proceeds north into high latitudes. In northern California, Tolowa and Yurok social structures are much like those of the typical nomadic hunter-gatherer; around the U.S.A.-Canada border, Coast Salish and, probably, Chinook social structures exhibit ranked classes within a bilateral kinship framework; further north in Canada, Bella Coola, Nootka and Kwakiutl social structures rest on cognatic descent groups and complex interpersonal and intergroup ranking; and around the Canada-Alaska border, Tsimshian, Tlingit and Haida social structures are based on matrilineal descent groups and preferential cross-cousin marriage systems (see Figure 1). I am

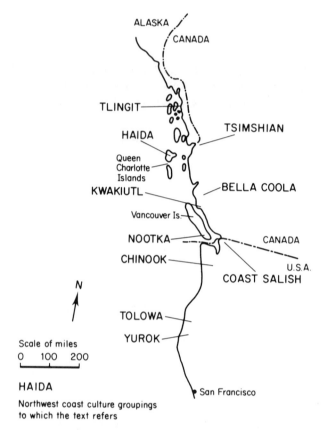

ALASKA

CANADA

TLINGIT

TSIMSHIAN

HAIDA

Queen
Charlotte
Islands

KWAKIUTL

BELLA COOLA

Vancouver Is.

NOOTKA

CANADA

CHINOOK

U.S.A.

COAST SALISH

N

TOLOWA

YUROK

Scale of miles
0   100   200

HAIDA

San Francisco

Northwest coast culture groupings
to which the text refers

Fig. 1. Sketch map of approximate locations of the main Northwest coast culture groupings.

strictly concerned with social structures held to exist immediately prior to enduring European contact.

This essay argues that ecological constraints may indeed be crucial to the variation in Northwest coast social structure. It is proposed that the constraints may be embodied in the notion that, proceeding north along the coast, there is increasing uncertainty in the production of food. It is hypothesised that the variation in social structure may be understood in relation to matters of productive uncertainty, as these are mediated through three independent 'basic structural notions', *viz.* notions of ownership, notions of rank, and notions of kinship; the first two notions differentiate Northwest coast social structures from those of most nomadic hunter-gatherers. It is argued that, particularly in the context of notions of ownership, increasing productive uncertainty means an increasing emphasis on the control of manpower. It is demonstrated that the increasing 'rigidity' in Northwest coast structures, as one proceeds from south to north, could relate to an increasing emphasis on the control of manpower. In a brief concluding section, it is suggested that the models for the cognatic descent and matrilineal descent structures proposed for the Northwest cost Indian may be applicable elsewhere.

Whatever explanatory methodology one employs, it is probably impossible to demonstrate conclusively that connections between ecological variation and the varying social structures of the Northwest coast do exist; the data available on pre-European ecology and society are almost certainly inadequate. The argument presented here is underlain by a transactionalist methodology, particularly as it is established in Holy (1976) and Stuchlik (1977). In this methodology, social structures are held to constitute ideologies, specifically statements about economic and political strategy. Ideologies are held to exist in so far as the strategies which they embody are pertinent to the actors' current purposes and situations, as the actors perceive these purposes and situations. In this essay, the postulate that the various statements about the control of manpower embodied in Northwest coast social structures relate to varying ecological situations rests on the assumption that the Indians are aware of these situations and of their consequences for production in the particular areas in which they live. I believe that the way in which the varying ecological situations will be presented makes the assumption plausible; however, in the absence of definitive evidence sustaining the assumption, the following ecological account of Northwest coast social structure is presented as a hypothesis.

The congruence between ecological variation and social structural variation on the Northwest coast suggests hypotheses about the origins of Northwest coast social structures. A concern with origins contrasts with the interests of most other writers on Northwest coast social systems (Suttles 1968a is a notable exception), who have focused mostly on the ecological function of specific institutions, or else on demonstrating a structural 'fit' among various institutions in specific social structures. The interest here is with why Northwest coast social structures have come to take their particular forms. It is assumed that the answer lies in the form of the social structure from which they may be supposed to have respectively emerged, this emergence having been mediated through changed ecological conditions. This approach is, of course, complementary to, and not mutually exclusive of, functionalist or structuralist interpretations of Northwest coast institutions. Indeed, since this paper tackles the whole range of Northwest coast societies and therefore cannot discuss the ethnographic nuances evident in the multitude of primary sources, I shall be explicitly building on the findings of these other approaches, particularly on Rosman and Rubel's excellent structural account (1971a; 1971b; 1971c; 1972).

Several writers have reasoned that the Northwest coast culture has its origins in some inland subarctic or Plateau culture, whose population spread to the coast and moved both north and south (Kroeber 1939: 29; Suttles 1968a: 105; cf. Spier 1930; Elmendorf 1977). Supposing that the original culture was much the same as that of the contemporary nomadic hunter-gatherers occupying the same area, I propose that we interpret (1) Coast Salish type social structures as emerging from a modified version of a typical nomadic hunter-gatherer type social structure; (2) the cognatic descent structures of the Nootka and Kwakiutl as emerging from a Coast Salish type social structure; (3) the matrilineal descent structures of the Tsimshian, Tlingit and Haida as emerging from a Kwakiutl type social structure.

In the following sections, I examine each of the elements in this argument in turn.

Ecological Variation of the Northwest Coast: Two Substantive Issues

A great deal has been written on the question of whether or not, or to what extent, particular human groups on the Northwest coast are subject to local, seasonal or annual fluctuations in the availability of productive resources (e.g. Suttles 1960; 1968a; Vayda 1961; Piddocke 1965; Drucker and Heizer 1967). Empirically, I am not competent to join this debate (cf. Hawthorn 1968). In any event, my concern is with evidence for differences in the availability of resources as one proceeds along the coast from south to north. Here, I seem to be on safer ground. Certainly, moving from the Coast Salish area northwards, one finds 'fewer species of edible bulbs and roots and less extensive areas where they can grow'; edible berries may also be less plentiful; 'we find fewer deer and perhaps less land game generally'; shellfish, while plentiful in the north are generally concentrated in more restricted areas; 'waterfowl are very likely less abundant'; fish and sea mammals are plentiful in all areas, but bad weather much more often impedes their exploitation in the northern regions. In sum, 'as compared to the . . . Coast Salish, the more northerly tribes rely on fewer kinds of plants and animals and get them at fewer places and for shorter times during the year, but in greater concentrations, and with consequent greater chance of failure' (quotations from Suttles

1968a: 102–3). When one focuses on resource groups, one supposes that such circumstances are exacerbated: normally numbering between thirty and one hundred people, these groups are small and occupy relatively restricted exploitative ranges. It is to the ecologic situation of such resource groups that I refer when I say that, proceeding along the coast from south or north, people are subject to an increasing productive uncertainty.

A second ecological issue I wish to consider in this section is the varying types of resources on the Northwest coast, and the implications this has for the sexual division of labour in exploitative production in the various societies. It seems to me that this has been a curiously neglected topic in ecologically-oriented analyses of the Northwest coast societies. The salient point here is that the proportion of food secured by women (i.e. plants and shellfish in the main) decreases as one proceeds from south to north; among the Tolowa probably around 50% of the diet is so secured; among some northern societies it may reduce to as low as 10% (Suttles 1968b: 61). Conversely, therefore, the proportion of food secured by men in hunting and fishing increases as one proceeds from south to north.

Notions of Ownership

Ownership connotes the exercise of exclusive rights in the disposal of resources or areas of territory. On the Northwest coast, ownership is exercised, varyingly, over *inter alia*, fishing sites, shellfish beds, berrying grounds, hunting areas, trade routes; rights of ownership are held, varyingly, by individuals and by groups. In the existence of notions of ownership, Northwest coast Indian cultures contrast with the cultures of most nomadic hunter-gatherers. To be sure, among nomadic hunter-gatherers, specific groups of people may be associated, as 'customary occupants', with specific tracts of territory; but in so far as rights of exploitation are normally not restricted to these groups, describing 'customary occupation' through 'ownership' and 'ownership-related' terms would seem to be misplaced (cf. Marshall 1976: 189–90).[2]

I shall not attempt to explain why Northwest coast hunter-gatherers assert ownership over resources. Clearly it relates to a number of factors, among them population density, size of exploitative groups, and the migratory behaviour of important animal resources. In this paper, I treat notions of ownership as a given feature of Northwest coast culture.

Notions of Rank

Rank connotes interpersonal deference, institutionalised in differential entitlements to tangible wealth, ceremonial objects and privileges. Among the more northern Northwest coast cultures, differential rank is importantly evident in 'orders of precedence' in potlatch ceremonial. In terms of social relations, rank is manifest varyingly among the Northwest coast cultures. In southern cultures it connotes simply interpersonal differentiations; in northern cultures it also connotes intergroup differentiations.

I believe that it may be shown that the existence of notions of rank in Northwest coast culture relates crucially to the fact that Northwest coast Indians entertain the exchange of food and wealth (e.g. Suttles 1960: 298), a procedure extremely uncommon among nomadic hunter-gatherers. Thus, among the Tolowa, rank is established through wealth, and wealth is established through dispensing surplus food in return for material goods (Gould 1966). However, as with ownership, I propose to treat the existence of notions of rank as a given factor in this analysis.

Notions of Kinship

Most social relationships in Northwest coast Indian culture are articulated in notions of kinship and affinity. Two points arise from this. First, kinship-based constructs are available for the conceptualisation of resource groups. Second, through ties of kinship and affinity people secure access to resources in neighbouring resource areas; this may or may not involve people's movement to these areas (Suttles 1960: 301; Rosman and Rubel 1971a: 26; Adams [1973] has made a comprehensive statement on this point). Through such kin connections, it may be said that all Northwest coast social systems are 'flexible', in so far as there are means for coping with differential pressures on areas of resources. Thus, in the cognatic descent systems, flexibility is achieved in the fact that the individual has connections with several descent groups (Harris 1968: 306; Adams 1973: 97); in the matrilineal systems, it is achieved through the individual exercising rights to the resources of non-matrilineal kin, particularly of members of the father's matrilineage (Adams 1973). A related point follows from this: invoking the fact that a particular social system is flexible cannot, in itself, explain why the system is found in a particular geographical area.

I should state here that, in this essay, I shall be drawing on a particular view of the existential status of descent constructs in social systems. This holds that, being part of the actors' conceptual apparatus, descent constructs should analytically be regarded as separable from actors' transactions 'on the ground' (Keesing 1971; Scheffler 1974); it holds therefore that descent ideologies may be regarded as, in effect, 'superimposed' upon social groupings on the ground (Sahlins 1965); and, in the realm of systems of preferential cross cousin marriage, it holds that what analysts and actors may conceptualise as preferential cross cousin marriage, reflects, on the ground, groups of people allied on the basis of some consistent transgenerational exchange of personnel (Schneider 1965).

The Hypothesis about Variation in Northwest Coast Social Structure: The Coast Salish

The hypothesis about variation in Northwest coast social structure relates, in the first instance, to the varying ways in which notions of ownership are operationalised in Northwest coast societies. In southern areas (e.g. Tolowa, Yurok), most individuals separately exercise rights of ownership over specific areas of wild resource (Kroeber and Barrett 1960: 3, 4, 115; Suttles 1968b: 64; Gould 1976: 71).[3] In central areas, resources are vested in specific individuals. Others secure rights of access either through manipulating kinship ties with these individuals (Chinock, Coast Salish), or through being members of these individuals' descent groups (Nootka) (Ray 1938: 56; Suttles 1968b: 65; Rosman and Rubel 1971a: 77–78). In northern areas, the ownership of resources is vested in corporate cognatic or matrilineal descent groups (Bella Coola, Kwakiutl, Tsimshian, Tlingit, Haida) (Drucker and Heizer 1967: 10; Rosman and Rubel 1971a: 111–13, 140).

The concern here is with the contrast between the Coast Salish (with whom we may probably lump the less well documented Chinook) and cultures to the south. Among the Coast Salish, the vesting of ownership in specific individuals may be related to relatively high productive uncertainties, and a relatively greater emphasis on male-oriented hunting and fishing in the securing of food. There is a parallel here with the Eskimo. Among the Eskimo, a good deal of productive equipment is owned only by a few individuals, the Alaskan Eskimo whale boat, for example (Spencer 1959: 175); these people have the special productive skills necessary to secure the substantial animal resources (skins, bones, etc.) needed for maintaining the equipment. In a similar way, one posits that the *development* of Coast Salish patterns of territorial ownership relates to the fact that, in the Coast Salish ecological situation, there is the facility for certain hunter/fishermen to evince a productive competence which fellow producers will recognise as importantly superior; these people attract a close stable following and thereby expand the territory which they, as individuals, control.

As one would expect, the operationalisation of notions of rank among the Coast Salish reflects Coast Salish patterns of territorial ownership. People who own major productive resources and their close kinsmen (who stand to inherit them) are reckoned as 'upper class'; others are 'commoners' or slaves (cf. Ray 1938: 56–7; Suttles 1960: 296). Unlike Northwest coast cultures to the north (from Nootka to Tlingit), Coast Salish do not acknowledge discriminations in rank, either among upper class people from a specific resource area or between people of different resource areas.

Discussion in the following sections takes the form of Coast Salish social structure as given. It supposes that the ideologies expressed in the more northern social structures may be interpreted as asserting increased emphasis on resource group control of manpower, for both economic purposes and warfare. It holds that the form these social structures take is predicated ultimately in the form of Coast Salish social structure. The hypothesis about structural variation on the Northwest coast embodies the assumption that, in a cultural context where ownership over areas of territory is asserted, increasing productive uncertainty means increasing emphasis on activities concerning the defence of territory and its wild resources, and the maximisation of the productivity of resources from that territory.

Data on warfare and slavery on the Northwest coast would appear to support these contentions. The further north one proceeds along the coast, warfare is more explicitly the concern of the resource group as a whole, and is more directly oriented towards matters of resource group economic expansion. Up to the Coast Salish, warfare is concerned mainly with matters of intergroup vengeance; among the Nootka, it occurs largely as a consequence of the *internal* political structure of the resource group, specifically its non-corporate character (Swadesh 1948: 93; Rosman and Rubel 1971a: 139); for the Kwakiutl, the literature disagrees as to whether or not warfare is mainly an economic concern (Drucker and Heizer 1967: 18; Rosman and Rubel 1971a: 139; Goldman 1975: 71). However, among the northern matrilineal societies, there is general agreement that resource groups engage in warfare mainly for economic plunder, slave taking, and, to a lesser extent, territorial expansion (e.g. Rosman and Rubel 1971a: 35). As far as slavery is concerned, there is a similar pattern. The further north one proceeds, the more slaves one finds (Ruyle 1973: 613—4). Moreover, up to the Coast Salish, discussion in the literature equivocates as to whether slaves constitute an economic asset or liability (Suttles 1968b: 65—6). One supposes that the economic importance of slaves in northern areas rests in the fact that, under conditions of high productive uncertainty, productive effort must be maximised when resources become available.[4] Here one notes the Northwest coast Indians' sophisticated techniques in food storage (Suttles 1968b: 63).

Cognatic Descent Structures on the Northwest Coast

The Nootka and Kwakiutl represent the main Northwest coast cultures where involvement in resource groups is conceptualised in terms of cognatic kinship connections with specific ancestors. The concern in this section is especially with the form of ranking within the descent group, and how this relates to the interests of the resource group.

For each descent group, there is a stock of relative ranked positions, the respective names and privileges of which a limited number of the group's membership will inherit. Inheritance of a ranked position is by primogeniture (however, women may not always exercise their privileges), and follows a basic 'conical clan' pattern, i.e. the top rank devolves by primogeniture from the founding ancestor's eldest child, the second rank similarly devolves from his second child, and so on (Drucker and Heizer 1967: 11—12). Thus contemporary second-born or younger children generally inherit lesser positions, or no position at all. In a cognatic structure, people can inherit positions from members of a number of descent groups; among the Nootka and Kwakiutl, people normally only enjoy the privileges of positions when they reside in the resource group with which a specific descent group is associated (e.g. Rosman and Rubel 1971a: 86). In addition to this system of ranking, there are, among the Nootka and Kwakiutl, a further set of positions which devolve from a man, especially a person of high rank, to his daughter's husband and daughter's sons (Goldman 1975: 66ff).

The chief difference between the Nootka and the Kwakiutl is that, among the Nootka, ownership of resources is vested in specific individuals (the highest ranked member of each descent group), while, among the Kwakiutl, ownership is vested in the resource group as a collectivity.

Discussion in the literature on the Nootka and Kwakiutl systems has focused, empirically and methodologically, on issues with which this analysis is not centrally concerned. Emphasis has been put particularly on the 'flexibility' of cognatic descent structures, i.e. on the facility, in a system of overlapping groups, for the redistribution of population in response to differential population pressures on resource areas, and for attracting manpower in periods of population decline, particularly in the postcontact era (Harris 1968: 306; Adams 1973: 97). Empirically, such discussion is basically about either people's manipulations of the cognatic descent structure or else the consequences of this structure for intergroup relations; in both instances the existence of the particular descent group structure is taken as given (cf. Rosman and Rubel 1971a: 159; Goldman 1975: 39). Moreover, the discussion does not normally attempt to explain the specific form Nootka and Kwakiutl ranking takes. Methodologically, the discussions usually embody functionalist assumptions; these focus on an institution's systemic consequences, not on the basis for its existence (cf. Orans 1975: 315—6). Further, as I indicated earlier, invoking a structure's flexibility is not likely to lead to explanations of its situation in a particular geographical area.

It is my view that Nootka and Kwakiutl social structures embody statements about *resource group* interest in manpower, the means to maximal productive and reproductive capacity. In reckoning descent cognatically, the involvement in a group of the widest range of kinsmen descendants is stressed as of critical importance. In the form of conical clan ranking, the crucial significance of children born early each generation, for the perpetuation of a stable membership core of a group, is asserted. Moreover, in extending positions to the daughter's husband and daughter's son, the importance to the group of children liable to be born elsewhere is emphasised. Since resource group exogamy and patrilocal residence are preferred (e.g. Rosman and Rubel 1971a: 149), the daughter's son's interest in the group is likely to be inhibited through his probable absence from it. Finally, among the more northerly Kwakiutl, making the resource group (*numaym*) corporate generally enhances members' interests in group concerns.

A group ideology that asserts interest in maximising the control of manpower is congruent with ecological conditions which embody high uncertainty in hunting and fishing and which engender strategies for production maximisation and territorial occupation and defence. Moreover, compared with Coast Salish social structure, Nootka and Kwakiutl social structures reflect a greater emphasis on the control of manpower; this corresponds with the greater productive uncertainty to which the more northerly peoples are subject. One may readily envisage a transformation of Coast Salish social structure into a Nootka/Kwakiutl social structure under conditions of increasing productive uncertainty. Simply through the introduction of a group-related notion of descent, a Coast Salish structure, where bilateral kinship provides individuals with access to many resource groups, transposes into a Nootka/Kwakiutl structure, where bilateral kinship enables resource groups to assert their interst in many individuals. In addition, as Suttles (1968a) has pointed out, Nootka/Kwakiutl ranking may be regarded as an elaboration and refinement of Coast Salish notions of rank.

Matrilineal Descent Structures on the Northwest Coast

Rosman and Rubel (1971a) have demonstrated that the social structures of the Tsimshian, Tlingit and Haida may be readily characterised in terms of prevailing anthropological models of systems of preferential cross cousin marriage. Resource groups are corporate matrilineages; a stock of names and privileges is associated with each lineage, though unlike the Kwakiutl and Nootka, succession is not through strict primogeniture (cf. Rosman and Rubel 1972). Residence, particularly among lineage members of high rank, is avunculocal.

In Tsimshian social structure, a series of lineages exchange women 'in a circle', i.e. there is a preference for MBD marriage (Rosman and Rubel 1971a: 14, 16). Though the participants in the alliance are formally equals, wife-taking lineages are in effect reckoned as superior in rank to wife-giving lineages. In Tlingit and Haida social structure, a series of lineages pass women in one direction in one generation, and return them in the opposite direction in the next, i.e. there is a

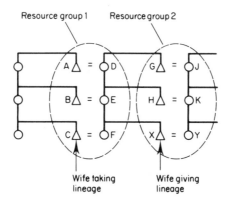

Fig. 2. Tsimshian: lineage diagram. A series of lineages may be linked in this manner, male members of some extreme right hand lineage eventually taking wives from the extreme left hand lineage. In all systems diagrammed in Figures 2–5 women never live in the resource area associated with their own matrilineage.

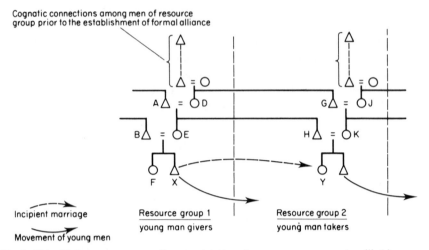

Cognatic connections among men of resource
group prior to the establishment of formal alliance

Incipient marriage

Movement of young men

Resource group 1
young man givers

Resource group 2
young man takers

Fig. 3. Tsimshian: resource group diagram. A series of resource groups may be allied in this manner, the people of some extreme left hand group eventually taking young men from the extreme right hand group.

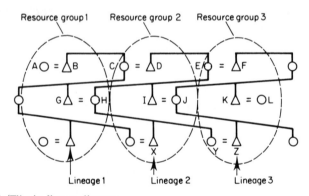

Fig. 4. Haida/Tlingit: lineage diagram.

a preference for FZD marriage. The lineage diagrams (Figures 2 and 4) represent these systems. Rosman and Rubel (1971a) have shown that the forms of both these types of social structure are recapitulated in patterns of inter-lineage property exchanges, particularly in the potlatch. Empirically, it is in alliances established among important lineages from neighbouring regions, or more specifically their high ranking representatives, that the structures are most directly evident.

Conceptualising these systems in terms of the exchange of women quite reasonably indicates how the systems operate. Moreover, the people themselves seem to see their social world in much the same way (Rosman and Rubel 1971a: 3, 13). However, I believe that this approach obscures some of the most crucial questions about North-west coast matrilineal structures. Why are corporate groups concep-tualised matrilineally? Why are alliances established through par-ticular modes of exchange? Why do Tsimshian reckon wife taking as superior to wife giving? Citing Homans and Schneider (1955) to explain Tlingit and Haida FZD marriage will not do (cf. Rosman and Rubel 1971a: 54–55). As Buchler and Selby (1968: 122–3) have indicated, the nature of alliance structures makes 'sentimental' explanations of cross cousin marriage systems inappropriate. More-over, such explanations will not account for Tsimshian MBD marriage. Concerning Tsimshian hypergamy, we note that Tsimshian interlineage relations are not those of caste or class (Rosman and Rubel 1971a: 12), the common basis to hypergamous marriage alliance (Leach 1961: 84). However, arguing that hypergamy is inevitable, since hypogamous arrangements mean that men of the matrilineage will be in a situation where they rank lower than their in-marrying wives, begs the question (cf. Rosman and Rubel 1971a: 184–5). The issue surely is: why are notions of rank expressed in the giving and receiving of women? In the following analysis I propose a model which copes with all these problems.

The model proposes, first, that we conceptualise these northern Northwest coast systems in terms of alliances between resource groups, rather than in terms of alliances between lineages (Rosman and Rubel 1971a: 15, 41) or even between local descent lines (Leach 1961: 56–63). It proposes also that we leave the traditional notion

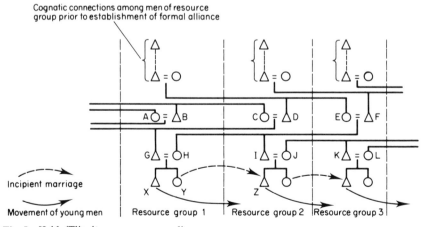

Fig. 5. Haida/Tlingit: resource group diagram.

that intergroup alliance is necessarily predicated on the exchange of women (i.e. in terms of the groups' respective reproductive potentials), and posits that, equally, it may be expressed in terms of the exchange of general productive potential (i.e. in the first instance in terms of the exchange of young men). This view concords with the hypothesis expressed in this paper, that Northwest coast social structures may be understood as being generated through strategies for the control of manpower.

The model for the generation of Northwest coast matrilineal systems is predicated on four facts. First, it is supposed that a series of resource groups is concerned in the establishment of alliance. Second, it is supposed that the alliance is contracted through the resource groups' mature men exchanging young male members of the group. (We may assume that the men in a resource group are not related matrilineally.) Third, it is supposed that, since the allocation of young men is being thus controlled, the allocation of young women will be controlled as well, as a complement to the allocation of young men. Fourth, it is supposed that intergroup transactions appropriately connote a formal inter-group equality.

Figures 3 and 5 present alternative ways of discharging these requirements. Since a series of resource groups are involved in a given alliance, both figures represent systems of generalised exchange. In Fig. 3 groups of mature men establish alliances, nominally 'in a circle'. They pass young men from group to group in circular fashion. Receiving groups allocate their daughters to in-marrying males. With this arrangement, specific inter-group transactions are relatively 'equal', especially given a broader context of circular alliance. The transactions are relatively equal in so far as pairs of groups are basically exchanging a man for a woman. Since these initial trans-actions connote a relative inter-group equality, they may be replicated in succeeding generations.

In Fig. 5 the resource groups' mature men similarly allocate young men, but they allocate young women in the same direction. Thus, man-receiving groups also receive women; they pass these women on, as 'son's wives', to the groups to which their 'sons' are destined. In this arrangement specific inter-group transactions are decidedly unequal — groups are sending men *and* women in the same direction. Inter-group equality is thus connoted through reversing the procedure in the succeeding generation. Since reversing the procedure means that an exact equality is expressed in inter-group transactions, this arrangement may proceed without, as a wider context, the groups establishing alliances in a circular fashion.

How do these procedures work out in kinship terms? Basically, we have posited two stages: first, the establishment of initial contracts; second, the perpetuation of alliance through succeeding generations. In the initial stage, we may propose the origin of matrifilial notions. Mature men exchanging young men must assert the exchangeability of these young men. Patrifilial or cognatic notions do not facilitate this, matrifilial notions do. Through stressing matrifilial connections, mature men differentiate themselves from the young men; the young men are defined as not having main connections with the mature men's group but rather with their mothers' groups.[5] In the second stage, we note the emergence of a pattern of avunculocal residence. We propose that, in a context of notions of matrifiliation, patterns of avunculocal residence facilitate the notion that resource group membership may be conceptualised in terms of matrilineal descent. Now, finally, we may read off from Figs. 3 and 5 who is marrying whom. In Fig. 3 we have MBD marriage (i.e. the Tsimshian system). In Fig. 5 we have FZD marriage (i.e. the Tlingit/Haida system).

This essay will not address generally the problem of inter-group ranking on the Northwest coast. However, the relationship between Tsimshian wife-taking and wife-giving lineages has been posed. In discussing this specific example, it is hoped that some general principles about inter-group ranking are being established, which may be elaborated elsewhere.

Empirically, Tsmishian lineages which enjoy high rank are those which control the greatest wealth and population (Rosman and Rubel 1971a: 17, 33) – from discussion earlier on the relationship between rank and wealth generally on the Northwest coast, we would expect this. However, our problem is with why rank difference is expressed in the form of wife giving and taking. In inter-regional alliances, which the foregoing model most clearly expresses, inter-lineage rank differentiation is more latent than actual, since the alliances are established among formal equals. However, within regions, important lineages are connected with less important lineages through wife taking and giving, which explicitly denotes their rank difference. As Rosman and Rubel demonstrate (1971a: 13, 16), the Tsimshian system is structurally analogous to *gumsa* Kachin (Leach 1961: 81–100).

I believe that the model shows why Tsimshian wife-takers rank higher than wife-givers, and therefore I hold that intra-regional lineage hierarchies are predicated on the existence of inter-regional alliances. The model invites the conceptualisation of alliance in terms of the exchange of young men, and that wife-taking lineages are regarded as young-man-givers, and wife-giving lineages are regarded as young-man-takers. We therefore reformulate the problem: why is giving young men reckoned as superior to receiving young men? To answer, we set the matrilineal models in the northern Northwest coast ecological context.

On the extreme northern Northwest coast, ecological conditions make for very high productive uncertainties, and thus recruitment into resource groups is carried out in a context of prevalent inter-group warfare and economic expansion. It is plausible to associate the Northwest coast matrilineal societies, as interpreted in the models, in terms of such political and economic conditions. Under conditions where competition for manpower is likely to be intense, the alliances may be regarded as agreements between important resource groups for the suspension of this competition.

Thus, the control of the young man's labour power is *both* the rationale for, and the medium of exchange in Northwest coast matrilineal alliance structures. From this, we appreciate why the Tsimshian young-man-giving lineage is reckoned as superior to the young-man-receiving lineage. As consistent donors of young men, the former group symbolically surrenders the crucial resource: the young man's labour power.

Compared with the Nootka and Kwakiutl social structures, the Northwest coast matrilineal alliance structures, as interpreted in this section, are consistent with relatively greater productive uncertainties. Moreover, we may envisage the transposition of the cognatic descent group structure into the matrilineal structures, under conditions of increased uncertainty. First, the mutual involvement of numbers of important resource groups in Nootka and Kwakiutl potlatch ceremonial – an occasion for political discussion as well as religious and economic display – predicated the emergence of systems of generalised exchange. Second, cognatic descent structures are the most plausible bases for the emergence of matrilineal systems as elaborated in the models. Cognatic ideologies readily admit the possibility that young men may be full members in more than one group, and may therefore be exchanged among groups. Finally, we may identify certain elements in cognatic structures which, under conditions of high productive un-certainty, may be expected to exacerbate conflicts over the control of manpower and so engender the formation of inter-group alliances. There is a latent contradiction in the Northwest coast cognatic structur As an ideology, it connotes an emphasis on maximising control of manpower. But 'on the ground' it readily facilitates people's movemen from one group to another, i.e. individuals invoke the structure in order to emphasise their simultaneous membership in several groups. The contradiction of course lies in the fact that, voluntarily changing one's group membership runs counter to the group's concern with retaining its members. That this contradiction indeed provokes inter-group conflict is suggested in Goldman's comment on the Kwakiutl:

> The genealogies testify to social disturbances created by the shifts of people from a father's to a mother's lineage. What the lineage feared most was loss. Intrusion was a gain (Goldman 1975: 39).

The replacement of cognatic descent structures with matrilineal alliance structures mitigates this particular problem and the inter-group conflicts that could arise from it. In the matrilineal systems, the redistribution of population may proceed under conditions of explicit local group alliance, i.e. under conditions where the possibility of warfare among groups is ideologically perpetually negated. Thus, on the northern Northwest coast, it is particularly to the father's matrilineage resource group that one goes for alternative economic support (Rosman and Rubel 1971a: 21).

This proposal for the emergence of matrilineal systems on the Northwest coast fits well the contrasts in intra-group reckoning of rank between the matrilineal and cognatic societies. I have argued that Kwakiutl inheritance of rank by primogeniture may be understood as a crucial evolutionary product in local group competition for manpower. Among the matrilineal societies, as I have mentioned, succession to rank positions is much more flexible (moreover names and privileges do not pass outside the descent group [Drucker and Heizer 1967: 70]). The model for matrilineal systems predicts this. Local groups in alliance have suspended competition for the control of manpower; on this account inheritance of rank by primogeniture becomes redundant.

General Remarks on Cognatic and Matrilineal Alliance Structures

Naturally, I have not been able to consider many aspects of Northwest coast social structure in this short paper; particularly I have not discussed inter-resource group relationships, except to a very limited extent. At this stage, however, I believe it may be useful to consider briefly the implications of the analysis presented here, for the interpretation of cognatic and matrilineal alliance structures outside the Northwest coast — noting that Rosman and Rubel (1971b; 1971c) have explicitly demonstrated structural parallels between Kwakiutl and Maori, and Haida and Trobriand. It may be stressed that the focus in this section is on the particular political and economic strategies these ideologies connote; we do not expect that Northwest coast type ecological conditions will be the only conditions through which such strategies are promoted.

Cognatic Structures

Here I consider Allen's provocative contribution (1971) to the discussion of the flexibility of patrilineal and cognatic structures in New Guinea and Melanesian land tenure. Citing data from the Nduindui of New Hebrides, Allen argues that, beyond certain levels of population pressure on available land, patrilineal structures will not effectively facilitate redistributions of population in response to differential pressures on specific areas of land; at this point, inherently more flexible cognatic structures necessarily emerge. The merit of Allen's contribution is its focus on the processes through which cognatic structures emerge and its identification of ecological preconditions for this emergence.

Following Meggit's suggestion (1965) that New Guinea resource groups tend to become more *de facto* patrilineal as pressure on land grows, Allen imagines what happens when this trend reaches its limit. Since resource groups are allowing agnates only to reside in their respective territories, people in hard pressed groups can get access to resources elsewhere through two ways: first, through warfare, and second through the emergence of ideologies in which, by right, non-agnates are allowed access to resource group territories. In both instances cognatic structures emerge: in the first instance, warfare means the accretion of members of the defeated group to the membership of the victor group and thus the emergence in a single resource group of substantial numbers of non-agnatic ties; in the second instance, cognatic structures are the most likely candidates for such a required ideology.

I am afraid that I find the various elements in this argument unrealistic. This may be because Allen's illustrative Nduindui data are not exactly apposite, since they refere to relative agnatic and cognatic emphases on group composition within a basic cognatic structure. In the first place, Allen does not account for the notion of descent in Melanesian cognatic structures; a focus on the requirement of structural flexibility fails to explain why a non-descent Coast Salish type bilateral structure does not emerge. In the second place, Allen seems not to appreciate that patrilineal ideologies tolerate extremely divergent 'on the ground' compositions (Meggitt's review of New Guinea highland societies indeed demonstrates this point); thus extensive warfare does not necessarily lead to their demise. Thirdly, one simply cannot envisage a resource group under generally hard-pressed conditions formulating an 'open house' cognatic ideology in order that others might more easily cope with their land problems.

The model for the generation of cognatic descent structures proposed for the Northwest coast presents a more realistic scenario here. We may posit that Melanesian cognatic structures embody statements about group maximisation of manpower and suggest that their emergence relates to situations of critical pressure on land, which make waging war, defending territory, or simply inter-group tension highly probable. Further evidence that Melanesian cognatic structures are about groups maintaining their integrity is suggested in the presence of certain restrictions on individual access to resource groups. Commenatry on the Nootka and Kwakiutl made the point that cognatic structures are readily manipulable by individuals seeking to transfer group membership, and that this, in effect, conflicts with the ideology's raison d'etre. From this point of view, social rules among, for example, the Kwaio and inland Nduindui, which state that only resident agnatic members of the resource group have automatic full rights to land, are of interest (Keesing 1971; Allen 1971: 11, 15–19). They assert that, while the group, through cognatic descent, may proclaim its interest in a maximal number of potential members, individuals may not equally readily secure access in a maximal number of groups. In times of colonial peace, as the raison d'etre of Melanesian cognatic structures diminishes, one would expect an increasing relative emphasis on rules restricting individual movement. Under cash cropping regimes, as among coastal Nduindui (Allen 1971: 18–20), this may reach its limit with the disappearance of corporate cognatic descent structures and the emergence of family estates.

Matrilineal Structures

Discussion in this paper suggests that the emergence of matrilineal alliance structures may be related to conditions where groups are in intense competition in respect of social activities for which the control of manpower is critical. Here, with regard to the Trobriand case, we note Brunton's suggestion (1975) that there is intense inter-district competition for the capture of *kula* valuables in this particular part of the *kula* ring and associate with it Rosman and Rubel's observation (1971b) that Trobriand chiefs contract inter-district alliances empirically of the Haida FZD marriage type. On more general concerns, the Northwest coast analysis suggests the following four points. First, matrilineal alliance structures are likely to be closely associated with neighbouring cognatic regimes. Second, the well known association of FZD marriage with matrilineal regimes (Homans and Schneider 1955) is to be understood in terms of a single social process, which generates, at the same time, this particular form of marriage and the matrilineal notion. Third, the matrilineal FZD alliance structure may be relatively uncommon (that is, compared with patrilineal MBD structures), because alliances proceeding through the exchange of young men are relatively uncommon. Fourth, matrifilial notions evident in matrilineal societies may have their origins in the evolution of the matrilineal structure; one thinks here, obviously, of the Trobriand 'denial of physiological paternity'.

Footnotes

I should like to thank colleagues and students in the Department of Social Anthropology, Queen's University of Belfast, for their many critical comments on a first draft of this paper.

1. For example, it may be argued that there may be nothing especially 'flexible' about nomadic hunter-gatherer social structure. I have suggested that, instead of focusing on why hunter-gatherers normally reckon kinship bilaterally, we should be asking why hunter-gatherers social structures mostly lack ideologies of descent (Riches n.d.).

2. In some nomadic hunter-gatherer societies, people 'ask permission' prior to using territory and resources in areas where they are not customary occupants (e.g. Marshall 1976: 190). I suggest that, in itself, this behaviour does not evidence ownership. Rather, it serves to provide the means through which the newcomers' generally peaceable intentions are expressed. But see Peterson's contribution to this volume.

3. The ethnographies also specify notions of collective ownership, for example in the form of villages exercising rights over stretches of beach (e.g. Kroeber and Barrett 1960: 125). However, accompanying discussion suggests that such 'ownership' might be better labelled as 'customary occupation' as spelled out earlier in this paper.

4. This view is inspired by Epstein's analysis (1967) of the economic implications of hereditary ties among Peasant farmers and Untouchable labourers in an Indian village.

5. This supposes that mothers do come from other groups, i.e. it supposes *de facto* group exogamy. On the Northwest coast, particularly among important resource groups, this nearly always seems to be the case. Moreover, in the cognatic descent societies, women are never fully incorporated into their husbands' resource groups (Rosman and Rubel 1971a: 145).

References

Adams, John 1973. *The Gitksan potlatch: population flux, resource ownership and reciprocity.* Toronto: Holt, Rinehart and Winston.

Allen, Michael 1971. Descent groups and ecology amongst the Nduindui, New Hebrides. In *Anthropology in Oceania: essays presented to Ian Hogbin*, L. Hiatt and C. Jayewardene (eds.). Sydney: Angus and Robertson.

Brunton, R. 1975. Why do the Trobriands have chiefs? *Man (N.S.)* 10, 544–58.

Buchler, I. and H. Selby 1968. *Kinship and social organisation*. London: Collier MacMillan.

Drucker, Philip and Robert Heizer 1967. *To make my name good: a reexamination of the southern Kwakiutl potlatch*. Berkeley: University of California Press.

Dunbar, M. 1968. *Ecological development in polar regions*. Englewood Cliffs, N.J.: Prentice Hall.

—— 1973. Stability and fragility in arctic ecosystems. *Arctic* 26, 179–186.

Elmendorf, W. 1977. Coastal and Interior Salish power concepts: a structural comparison. *Arctic Anthropology* 14, 64–76.

Epstein, T.S. 1967. Productive efficiency and customary systems of rewards in rural south India. In *Themes in economic anthropology* (A.S.A. monogr. 6), R. Firth (ed.). London: Tavistock.

Goldman, Irving 1975. *The mouth of heaven: an introduction to Kwakiutl religious thought*. New York: John Wiley.

Gould, R. 1966. The wealth quest among the Tolowa Indians of north-western California. *Proc. Am. Philosophical Soc.* 110, 67–89.

—— 1976. Ecology and adaptive response among the Tolowa Indians of northwestern California. In *Native Californians: a theoretical perspective*, L.J. Bean and T.C. Blackburn (eds.). Ramona, California: Ballena.

Guemple, Lee 1972. Eskimo band organisation and the 'DP' camp hypothesis. *Arctic Anthropology* 9, 80–112.

Harris, Marvin 1968. *The rise of anthropological theory*. New York: Crowell.

Hawthorn, H. 1968. Review of P. Drucker and R. Heizer, To make my name good. *Man (N.S.)* 3, 678–80.

Helm, June 1968. Bilaterality in the socio-territorial organisation of the arctic drainage Dene. *Ethnology* 4, 361–385.

Holy, Ladislav 1976. Knowledge and behaviour. In *Knowledge and behaviour*, L. Holy (ed.). Belfast: Queen's University.

Homans, G. and D. Schneider 1955. *Marriage, authority and final causes*. Glencoe, Ill.: Free Press.

Keesing, Roger 1971. Descent, residence and cultural codes. In *Anthropology in Oceania: essays presented to Ian Hogbin*, L.R. Hiatt and C. Jayewardene (eds.). Sydney: Angus and Robertson.

Kroeber, A.L. 1939. *Cultural and natural areas of native north America*. Berkeley: University of California Press.

Kroeber, A.L. and S. Barrett 1960. Fishing among the Indians of northwestern California. *Antrhopological Records* 21, 1–210. Berkeley: University of California Press.

Leach, E.R. 1961. *Rethinking anthropology*. London: Athlone Press.

Marshall, Lorna 1976. *The !Kung of Nyae Nyae*. Cambridge, Mass.: Harvard University Press.

Meggitt, M. 1965. *The lineage system of the Mae-Enga of New Guinea.* London: Oliver and Boyd.

Orans, Martin 1975. Domesticating the functional dragon: an analysis of Piddocke's potlatch. *American Anthropologist* 77, 312–28.

Piddocke, Stuart 1965. The potlatch system of the southern Kwakiutl. *Southwest J. Anthrop.* 21, 244–64.

Ray, Verne 1938. *Lower Chinook ethnographic notes.* Seattle: University of Washington Publications in Anthropology.

Riches, D. n.d. Territorial ideology and bilaterality among nomadic hunter-gatherers. Paper presented at conference on Hunter-gatherers. Paris, June 1978.

Riddington, R. 1969. Kin categories versus kin groups: a two section system without sections. *Ethnology* 8, 480–7.

Rosman, Abraham and Paula Rubel 1971a. *Feasting with mine enemy: rank and exchange among Northwest Coast societies.* New York: Columbia University Press.

—— 1971b. Potlatch and *sagali*: the structure of exchange in Haida and Trobriand societies. *Trans. N.Y. Acad. Sci.* 32, 734–42.

—— 1971c. Potlatch and *hakari*: an analysis of Maori society in terms of the potlatch model. *Man (N.S.)* 6, 660–73.

—— 1972. The potlatch: a structural analysis. *Am. Anthrop.* 74, 658–71.

Ruyle, Eugene 1973. Slavery, surplus and stratification on the Northwest Coast. *Current Anthrop.* 14, 603–30.

Sahlins, Marshall 1965. On the ideology and composition of descent groups. *Man (O.S.)* 65, 104–7.

Scheffler, Harold 1974. Kinship, descent and alliance. In *Handbook of social and cultural anthropology*, John J. Honigmann (ed.). New York: Rand McNally.

Schneider, D. 1965. Some muddles in the models: or, How the system really works. In *The relevance of models for social anthropology*, M. Banton (ed.) (A.S.A. monogr. 1). London: Tavistock.

Spencer, Robert 1959. *The north Alaskan Eskimo: a study in ecology and society.* Washington: Bureau of American Ethnology.

Spier, L. 1930. *Klamath Ethnography.* Berkeley: University of California Publications in Anthropology.

Stuchlik, M. 1977. Goals and behaviour. In *Goals and behaviour*, M. Stuchlik (ed.). Belfast: Queen's University.

Suttles, Wayne 1960. Affinal ties, subsistence, and prestige among the Coast Salish. *Am. Anthrop.* 62, 296–305.

—— 1968a. Variation in habitat and culture on the Northwest Coast. In *Man in adaptation*, Y. Cohen (ed.), vol. 2. Chicago: Aldine.

—— 1968b. Coping with abundance: subsistence on the Northwest Coast. In *Man the hunter*, R. Lee and I. DeVore (eds.). Chicago: Aldine.

Swadesh, M. 1948. Motivations in Nootka warfare. *Southwest J. Anthrop.* 4, 76–93.

Vayda, Andrew P. A reexamination of Northwest Coast economic systems. *Trans. N.Y. Acad. of Sci.* 23, 618–24.

ECOLOGICAL STABILITY AND INTENSIVE FISH PRODUCTION: THE CASE OF THE LIBINZA PEOPLE OF THE MIDDLE NGIRI (ZAIRE)

PIERRE VAN LEYNSEELE

Introduction

The theoretical purpose of this paper is to contribute to the development of a formal theory which would relate a given environment – an 'ecological system' – and the population which occupies it – 'a social system'. More specifically, I shall examine the relations which exist within the former system and indicate only their more obvious connections with the social system.

The vast area between the Congo and Ubangi rivers, from 2° 30' N to their confluence, is occupied by more than thirty discrete groups. They can be reduced to four distinct clusters by a complex of criteria including regional peculiarities of the environment, specialisation in production, characteristics of the social structure and of linguistic and historical affiliation. There are several transitional groups between the various clusters. Transfers of people and commercial exchanges between the groups were constant. The Libinza were a numerous group, representative of the dominant cluster. Their system of intensive production has been progressively abandoned, so for the last four years only the hereditary rights to unworked sites remain.

For present purposes, the 'ecological system' is conceived as being constituted by three sub-systems: the ecosystem, the cognitive system, and the system of exploitation of the environment. The analysis of the ecosystem will reveal certain fundamental principles relating to ecological factors and their variations, which will necessarily recur as components of the other sub-systems. A close relation must exist between the ecosystem and the system of exploitation, and a similar relation must reveal itself at the level of the cognitive system, that is, at the level of the people's own knowledge of their environment.

The point of departure is an analysis of the ecosystem. But this cannot be restricted to the immediate environment occupied by a given population. It must necessarily range over the ecological processes of the entire zone of which the niche is only a part. Indeed it is essential to establish from the start the various stages of the spontaneous ecological processes, the succession of different ecosystems with their associated plant-life, whereby we may characterise the entire ecological cycle in its natural state. The general ecological principles which characterise the entire zone, and the variations in ecological factors relevant to each stage of the cycle must be defined. It will then be possible to establish to which of these stages the niche corresponds, and further to define the ways in which it differs as a consequence of the relative importance of ecological factors and of new factors, substituted or introduced, which produce explicable reactions. The significance of the human factor thus appears with reference to the spontaneous factors. If, as with the Libinza fisher-men, the niche is densely and permanently inhabited, human action aims to intervene to stabilise the ecosystem at a favourable stage, from the point of view of intensive and permanent exploitation. We may thus expect the analysis of the niche to reveal evidence of stabilisation consequent on an ancient and continued human occupation.

The understanding which people have of their actual environment cannot be established solely by the enumeration of the terms by which they refer to environmental elements and significant seasonal changes. Such an enumeration reflects a static view. This view is without value if one wishes to grasp the manner in which people com-prehend ecological processes, a comprehension which they must have for the intensive and continued exploitation of their ecological niche. Only an analysis of the distinctive components of the terms as they relate to the ecological processes can verify the people's own under-standing of the general ecological principles, the specific factors which affect the niche and also the seasonal variations. An understanding of the niche alone is inadequate, since it is subordinate to the dominant factors which characterise the zone as a whole. Moreover, if we are dealing with a niche in which production is specialised, the population necessarily depends in part upon external resources, which may become more significant than the local resources for at least a portion of the population.

The exploitation of the environment implies the utilisation of favourable ecological factors while maintaining the environment's ability to regenerate. It supposes knowledge and constant obser-vation of those signs which warn of the danger of over-exploitation, at which moment restrictive measures must be put into operation.

An analysis of an ecological system in this manner allows precision as to the limits imposed on a 'social system'. Competition develops within a social group for these sources in the environment which assure regular and abundant yields. A cleavage is established within the group between those members who possess rights over the most profitable sections of the environment and those who do not. In the Libinza case, this cleavage introduces opposed tendencies within the social group, one providing for its stability and continuity and the other oriented towards the outside. The limits of the exploitation require a constant adjustment of the numerical composition of the group to the available resources.

The Main Ecological Factors

The area of the Congo-Ubangi confluence is entirely dominated by the regimen of these two great rivers. The River Ngiri, a tributary of the Ubangi, virtually bisects this triangle. For the greatest part of the year, the sheet of water extends over a large area of the interior of the forest on each side of these rivers (and indeed downstream of their confluence). The annual variation of the water level averages 2.80 metres for the Congo and four metres for the Ubangi. These considerable variations completely condition the ecological processes. Precipitation is stable throughout the year and locally is only a minor influence on the water level.

Sandbanks appear during the low water season as a result of a shift in the current or of a minor change in the course of the river bed. These sandbanks are colonised by *Vossia cuspidata*, a grass with stems several metres long, allowing a period of total immersion. This grass has strong rhizomatic roots, which stabilise the alluvial deposits, while the large leaves slow down the current. The process of raising the banks prepares the site to receive a pioneer *Alchornea cordifolia* association, a dense shrub three to four metres high which ensures considerable alluvial deposits. At this stage, what was originally a totally flat sandbank has become a floating meadow with raised banks and a shallow internal terrace. This general profile is maintained. Whenever the alluvial bank is completely raised along any straight stretch of the main course of the river, or along the many straight channels between the islands, it is occupied by the riparian *Uapaca heudelotii.* The individuals of this forest form a uninterrupted fringe 25 to 35 metres high. The association with *Oubanguia africana* and *Guibourtia demeusei* is typical of alluvial banks which are flooded not more than three or four months a year. These conditions are frequently found close to the main course of the River Congo. The upper stories of this multi-canopied forest is 35 to 40 metres high; a lower denser one is 20 to 25 metres high while there is a level of extremely compact trees between 8 and 15 metres. This association represents a mature stage of the forest and entails a stabilisation of the environment and very probably the interruption of the alluvial process. A forest dominated by *Gilbertiodendron ogoouense* appears in a few scattered places on narrow strips of dry land. This forest can withstand occasional temporary floods.

Where the banks are occupied by a mature stage of the forest, they dominate the river like a cliff and their foundations are undercut by the current at the rate of about 0.30 metres a year. Other forms of erosion, representing all the stages of the seasonal cycle, continually modify the course of these rivers, whose minor beds are sometimes displaced by an exceptional flood.

The Ecosystem of the Libinza

The 36 Libinza villages are found along the River Ngiri, from a point about latitude two degrees north, where two branches of the river rejoin, to about 70 kms below this confluence. The land surface is fractionally higher above river level downstream, but the same ecosystem is dominant throughout Libinza country.

Despite the numerous natural channels through the flooded forest between the Ngiri river and the Congo, the regimen of the waters in the Ngiri depends on the level of the Ubangi. The Ubangi's regimen is characterised by considerably more variation than the Congo, on average four metres and by a very rapid lowering, occurring between the middle of November and the middle of February.

In this section of the Ngiri, only the first two stages of bank formation are found, namely the growth of grasses and, in places on the banks, the association with *Alchornea cordifolia*. The practice of annual burning in May, before the period of flood, has inhibited the development of later phases. *Vossia cuspidata* has been replaced everywhere, except on the most recent banks, by *Echinochloa pyramidalis*, a grass which resists the annual burning well, and by *Jardinea* in those areas of the meadow furthest from the course of the river.

As a result of the brake on the process of bank-building, the river has only one minor channel flowing at lower water. During most of the year the banks are submerged, the waters flood all the meadows and only the compact mass of the tips of the grasses emerges. The countryside appears as a long corridor of floating meadow, four to six kilometres wide, bounded by the drowned forest on each side and containing the serpentine meanders of the main course of the river

with its extensions and pools (see Figure 1). The soil is only superficially fixed by the roots of the grasses and the erosion of the con-
cave banks only accentuates the curves of the meanders. During the
flood season, the current flows over the banks and finally links
adjacent bends by a new straight course. The old course progressively
becomes a minor bed and an extension of the new principal course.
Grass occupies the alluvial terraces delimited by the current and the
old beds of the river. These are progressively invaded by grasses,
which speed up deposits of alluvium, so that they are levelled. These
old beds become pools joined to the main course by a channel or, if
they are distant, completely surrounded by grass. [1]

Fig. 1. The countryside appears as a long corridor of floating meadow bounded by the
drowned forest on each side and containing the serpentine meanders of the main course of
of the river with its extensions and pools.

If the practice of annual burning was to cease, this ecosystem
would be invaded by forest in about a century. Human interference
is essential for its stabilisation. The virtually absolute domination of
the environment by *Echinochloa pyramidalis* and the absence of
natural mounds higher than the level of the banks suggest continuous
and ancient human occupation.

The Categorisation of the Environment

An analysis of those terms in the Libinza language referring to the

categories of the environment indicates a close correspondance to the
features of the natural environment and to the ecological process as
they were summarised above.

Loi describes the principal course of the river. It is not *per se*
obvious that a term should be reserved purely for that part of the
river contained between the levees of its minor bed which is never
completely dried out. In this section the Ngiri is a labyrinth. Often
the principal course is only about ten metres wide in places where
the extensions are more than a hundred metres wide and ten
kilometres long and receive a share of the current. The current is
only shown by a rippling wave of grasses on the convex banks,
whereas the surface of the water is just like a mirror. Because
of this, the fishermen themselves may make mistakes when they
travel since they use shortcuts which cut across the bends of the
meanders. In addition, in certain places where the main course may
not yet be stabilised, different routes may be followed according to
the season. *Loi* is a concept whose definition cannot be understood
except by an analysis of ecological processes. It is useful because it
is the point of reference for the observation of the gentle shifts in
the speed of the current, which have predictive value.

Mungala is an extension of the principal course, a part of an old
bed in which there is sometimes some of the current, or alternatively a new bed in process of formation.

Moluka (plural *miluka* or *botena* from -*ten*- 'cut') is a narrow
straight channel which cuts across the floating meadow. These small
channels are kept open by the constant passage of the fishermen, who
push aside the grass and pull it up by the roots. An examination of
aerial photographs reveals the straight track of the *miluka* which
converge on the inhabited spots and join them to the forest. In the
flooded forest, *moluka* is a natural drainage channel kept open by
regular passage and by the removal of fallen trees and other
obstacles. Formerly, several of these channels joined the River Congo
to the Libinza section of the Ngiri.

Molala is a small natural channel, a vestige of an old bed of the
Ngiri, partially levelled, which ensures the exit of water from a pool
at the end of the flood. This small channel often joins several pools
and flows into a *mungala.* The *miluka* sometimes include a *molala*
for part of their course.

Ntena is a drowned terrace occupied by *Echinochloa pyramidalis*
and situated between the present bed and an older bed of the river.
The edge of this terrace is bordered by the levees of the banks.

Modziba is a vast pool, sometimes more than a kilometre at its greatest width, joined by a channel (*molala*) to the main course of the river or to an extension. *Edziba* is a shallower ox-bow lake which is never completely dry at lower water. Sometimes several *edziba* are joined by a *molala*, but they are never linked to the main course of the river. *Limpoko* is a small shallow pool in the glades on the edge of the forest. It dries out completely during low water.

This categorisation coincides with the elements of the environment as they result from ecological processes. The distinction is made between natural and man-made elements as seen in the term *moluka*, the only artificial element. There is also a difference in function, as a *moluka* serves for communication while a *molala* establishes a link between the pools. However, in practice, the distinction is often hard to establish.

The categorisation also establishes a distinction between the different phases of the ecological process. *Loi* is the central concept, corresponding exactly to the notion of minor bed of the river (see Figure 2). The distinction with *mungala* is often subtle, the sole criterion being that during low water, all, or at least most of the current is in the *loi*. The different pools are categorised according to whether or not they are directly or indirectly connected to the main course or whether the connecting channel has been levelled. *Limpoko*

Fig. 2. *Loi* is the central concept, corresponding exactly to the notion of the minor bed of the river.

corresponds to an even older stage in the ecological process. Other indications, relating to depth, relative distance from the main course and sensitivity to fluctuations in the level of the river, derive from this classification.[2] *Ntena* is the only element in the environment completely covered by grass and is thus the final point of the ecological process.

It should be noted that no category groups elements in a manner irrelevant to the ecological process. The terms *loi* and *mungala* apply to general ecological processes. The knowledge of this relationship can be used throughout the whole zone.

The Classification of the Seasons

The state of the waters is the only criterion used to classify the seasons. The regimen of the Ngiri depends on the Ubangi for its major variations. The regimen of the latter, which is subtropical for most of its course, is characterised above all by a remarkable regularity, both as to the size of the shifts and the date at which they occur. However, precipitation in the forest upstream and the enormous extent of the drainage in the basin of the Ngiri river cause oscillations in its level and put it out of phase with the Ubangi. This therefore brings about periods during which the level of the Ngiri waters are unstable, before the flood or the ebb of the Ubangi makes itself felt. These periods are critical for the behaviour of the fish and for the choice of the fishing methods. In contrast, those periods in which the same conditions prevail for a known length of time are best for fishing, since the fisherman knows he can use the most efficient method for a given period. Knowledge of the predictability of the seasonal variations allows fishing operations to be planned and reduces the effect of chance.

Libukanaka is a short period in late November or early December during which the waters are at their highest. The speed of the current varies and the level oscillates from day to day. At the end of this short season, the ebb already operating in the Ubangi is felt. From then on, during the *moya* season, from December to mid February, the water recedes rapidly from the high point of the flood to virtually the lowest, in all a drop of about three metres. At the end of this season, the water is held in the minor bed of the river and in those of its extensions which are free of grass. The current decelerates. During the *elanga* season, until the beginning of April, the waters continue to drop slowly to their lowest level. Sand banks appear everywhere in the course of the river. The meadows, the forest and some pools dry up. The banks appear above the water. From early April to late

May, during the *motinda* season, the level increases slowly following the beginning of the flood in the Ubangi. Grass-burnings are organised then. The month of June, *moboma*, is the most difficult season. The flood of the Ubangi makes itself felt and the waters rise regularly. Fishing brings virtually nothing. From July to early August, during *mpuma*, the current accelerates, the waters rise more rapidly, but with temporary ebbs. From the end of August, during the *mpela* season, the flood is regular and continues until the end of November.

Conditions are regular and predictable during the progressive flood (*mpela*) from August to late November and again during the ebb (*moya*) from December to February which slows down until early April. For the rest of the year, the unstable conditions are not favourable for fishing. Every shift in the state of the waters is critical for the success of the fishing. The changes in the current in the main course of the river are expressed by the terms meaning push, pull, stop, accelerate and are now explained by variations in the Ubangi. Thus, during *moboma*, the waters of this river 'push' into the Ngiri, while during *moya*, the Ubangi 'pulls' the water from the Ngiri. The phenomenon is explained as having external causes.

The analysis of the Libinza categorisation of the environment shows a remarkable agreement between Libinza terms and the elements of the environment as they relate to ecological process. Knowledge of this environment is a precondition for the efficient exploitation of the habitat via intensive fishing. In addition, human intervention through the annual fires constitutes an ecological factor which inhibits the final phases of the soil building process and of the associations of plants characteristic throughout the area. Without human interference, the ecosystem of aquatic meadows with the expansion of the main course, a very favourable setting for intensive fishing, would give way to the forest.

The seasonal cycle is categorised by reference to the level of the water and the state of the current in the main course of the river. Knowledge of the configuration of the environment and the way in which water penetrates it, in conjunction with that of the seasonal cycles, allows precise knowledge of the state of the water at any moment. This information has predictive value. This knowledge is indispensable to the organisation of the fishing campaign, to the management of established fisheries and to the practice of individual fishing.[3]

The Libinza perceive the main course of the river as the dominant element in the environment. It conditions both the relatively static aspect of the ecological process — the modifications in the configuration of the environment, and the dynamic aspect — the continual variations in the state of the water.[4]

This environment, which at first sight seems hostile, is in fact much more hospitable than the surrounding forest, with its mosquitoes and tsetse flies. The mosquitos cannot breed in the acid water (pH 3.6) of the Ngiri and the floating meadows are not suitable for tsetse. Formerly, those who had sleeping sickness were transferred to islets separated from the community.

The Seasonal Movements of the Fish

From July (*mpuma*) when the current begins to speed up and the waters invade the terraces, the fish start moving and hunt along the banks for the passages into the floating meadows which act as spawning grounds. The separated stems of the grasses allow a simple passage, and the leaves floating on the surface give good protection against the sun. The fish find abundant food in the floating meadows.

During the ebb, the fish leave the floating meadows to collect in the pools, the extensions of the river and its main course where they stay confined during the season of low water until the flood starts in July.

As a result of the great oscillation in the water level and the rapid ebb at the beginning of the year, the environment of the Libinza acts as an immense fish trap. The principle consists of letting the fish enter the flooded meadows during high water, when they are attracted by abundant food and good spawning grounds, and keeping them there during the ebb when they try to take refuge in the deepest water. Every year the fish spread out evenly over the flooded meadows.

The Transformation and Utilisation of the Environment

Before the colonial period, the drowned meadows of the Ngiri were one vast fishery.[5] Around the villages, the possibilities of the environment were intensively exploited but without major alterations to the natural configuration of the area. The natural levees at the edge of the terraces were raised, passages were left open to allow the passage of water and pools were closed off with wicker work. E. Wilverth, one of the first Europeans to reach the Libinza reported (Goffin 1909: 32):

> ... nous nous trouvâmes dans un chenal de 4 mètres environ de large [a *molala*] et de distance en distance, nous rencontrions d'immenses clayonnages fermant le passage; une grande claie en fibres tressées formant tamis était retenue par quatre pieux et pouvait au moyen de lianes, être élevée ou abaissée à volonté; élevée le passage était libre; abaissée, le

poisson ne trouvait plus d'issue et se faisait fatalement capturer par les
natifs; à d'autres places il y avait de véritables digues en terre et en
branches d'arbres, s'élevant jusqu'à 2 métres de hauteur et retenant les
eaux. Donc, tout un système de pièges parfaitement compris et disposés
sur environ 2 à 3 lieues de chemin.[6]

The fish thus retained in the *ntena* and the pools were so concentrated that it was necessary to feed them. The possibilities for smoking
and immediate consumption needs determined the tempo of catching
these fish. Thus, this intensive fishing was a beginning of fish-farming.
Finally, during low water (*elanga*), the remaining fish in the slowly
drying sheets of water were scooped up in basketwork shovels during
communal fishing sessions.

In addition to this method of rationally using the configuration of
the environment and the variation in the water level to catch fish,
fishing with a variety of implements such as fish-baskets, nets, traps,
lines and prongs, was always practised in open waters as well as in
those *ntena* not set aside for intensive fishing. This was done by the
owner of the implements, aided if necessary by one or more of his
relatives. The success of this fishing depended on the judicious use of
whichever method was appropriate to the state of the water at any
given moment.

While the first method allowed superabundant production, the
second provided the needs of consumption while the waters were
rising late in the year and falling early in the following one. During
low water, however, only methods relying largely on chance
provided occasional catches.

Settlements are built on islets twenty to thirty metres across at
their broadest. These islets were formed artificially by building up
the banks along the main course of the river above the level of the
floods. The profile of the banks, sharp on the river side and then
sloping away from it, explains the general pattern characteristic of
the zone where settlements are built longitudinally along the line of
the river banks (see Figure 3). A Libinza village is formed by a certain
number of these islets, each one the property of a family group. As
the main course of the river is never totally dry, the fishers live along
a permanently open line of communication. *Libongo* signifies the
raised bank along the river in front of the huts but also, by extension,
the area for the exchange of produce.

The Bases of Social Groups

From the above, it should be clear that the possession of fisheries and
pools gives a considerable advantage in production. In principle, the

Fig. 3. The profile of the banks explains the general pattern characteristic of the zone where settlements are built longitudinally along the line of the river banks.

first occupant of an area has proprietary rights, which become exclusive as soon as the area shows evidence of being worked and exploited. The same rights of first occupancy applied to the transformation of the banks and the erection of islets. All those elements of the environment which were not being exploited or which could not be — thus *loi* and *mungala* — were open to all members of the village community.

The fisheries, the pools, and the islets were the private property of family groups controlled by the eldest of several brothers, who succeeded each other in order of age. Each of them occupied an islet or separate section of one with his family and dependants. In more numerous groups, the authority of the eldest, the *mata*, extended over his patrilateral parallel and cross cousins who could all legitimise their claims to the group's property. Outside this group, other members could belong to the same community by virtue of a choice of residence, a contract of clientage or as slaves, but without exercising a right to the property of the group. The numerical composition of the group depended on the size of its territory and on its prosperity. The basis of the social group was residential and the agnatic core effectively provided for its permanence. Impoverished groups or ones dying out for lack of descendants could always negotiate alliances or treaties in exchange for pools or islets. The territory could change hands and the kinship group effectively disappear.

The kin group exploited its fisheries in common and shared the produce in equal parts by family. The other members of the group, who did not have property rights, were allowed to participate in the exploitation, for which they received their share. But this participation was thrown into question when it was no longer justified as a result of a decrease in production.

Each family was free to dispose of that share of the production that it received. Just as the fishing instruments were the inalienable property of the fisherman, so the yields of the individual fishing remained family property. Fish was practically the only negotiable product and was needed not only for immediate consumption but also as a means of exchange to acquire necessary goods and produce that were not locally produced. Apart from a very small amount of horticultural produce and maybe pottery, fish was the only product which could assure the subsistence of the family and represented the necessary reserve for those seasons in which fishing yields were low.

The Limits of Production

The resources of the flooded meadows were intensively exploited and supported a dense human population. The density of the population was still 13–14 per square mile in the late 1950s (Wilmet 1974). All modern and old villages are from two to four kilometres from each other. In 1896 Wilverth (1896: 575) had noted: ' . . . je pus me rendre compte de la richesse du pays et de l'immense densité de la population'.

The Libinza methods prevented overfishing.[7] Each year the managed fisheries provided a certain proportion, determined by their area, of the total resources of the environment. This amount of fish was relatively stable. At low water (*elanga*), all the fish which had escaped capture remained in the open waters where they were difficult to catch for several months until, with the beginning of the floods, they started to move again. If the number of these fish decreased too much, a series of measures ended by limiting the exploitation of the total resources.

When, as a result of too great a population density, a locality was overfished, all the fishermen noticed it as individual catches became noticeably less. The owners of the fisheries then restricted participation in their exploitation. To avoid finding themselves witout reserves during the difficult season, those members of local groups without proprietary rights preferred to anticipate this and settle elsewhere.

The limited space available on the islets had the same effect on total local population numbers. An extension of the islets was only envisaged if a parallel extension of the fisheries was possible or if the other fish resources were adequate. There was thus a relation between the resources in fish and the habitable space available in each village community.

The structures of kinship and the modes of alliance offered an individual or group an extensive choice of possible affiliations to other groups linked either by kinship or by relations or clientage. Movement by individuals or groups was frequent and was motivated not only by necessity but also when better opportunities were seen elsewhere. Throughout these moves the individual retained his inalienable rights in his group of origin, which could always be claimed.

Orientation Towards the Outside

Two opposed tendencies characterise Libinza society. The agnatic core of the local group attached to its territory tended to be maintained in one place over the generations while the affiliated members, the clients, felt free to take any better opportunity of residence when it arose.

However, the whole community depended on its external relations to dispose of its fish and to acquire in exchange food or indispensable goods like dugouts and iron. Since the villages were situated along the main course of the river, transport of heavy and bulky goods over long distances was simple at all seasons and no chance was missed to exchange news or produce with a traveller.

Smoked fish was prepared for sale by an age-old procedure and then packed in baskets of standard size. Fish preserved in this way would stay edible for several months. Part of this fish was kept for consumption during the difficult season and the rest was sold. Each kin group maintained its own network of commercial relations and extended it by a policy of deliberately dispersed alliances. A multiple-purpose iron blade money was in circulation. In each family, the first wife was in charge of the operation of selling fish and buying food. The money could be used for the purchase of food but also as a reserve fund which could only grow with the development of commercial operations. Family savings were invested in the purchase of dugouts or in new matrimonial alliances which would themselves increase the capacity for production and commerce.

Smoked fish was in demand among the agricultural populations on the edge of the marshes and commanded a premium in exchange for the products of dry land. The Libinza, specialising in intensive fish production, benefitted from this situation, despite a series of inter-mediaries taking profits before sale in the dry land markets. The use of money and the regional specialisation of production diversified and stimulated these exchanges. During the last half of the nineteenth century, the Libinza had opened a direct link to the River Congo and formed the final link to the interior in this sector of what has been called the 'Great Congo Trade' (Vansina 1973: Ch. X). Thus they had access to imported valuables and prestige goods which they traded at a great profit along their own trading network to the edge of their territory. The intensification of commercial activities finally led to a concentration of riches and political authority in the hands of the heads of kinship groups who were particularly able to gather around them a numerous clientele.

Conclusion

In comparing a niche which shows evidence of significant human manipulation or intervention with the general spontaneous ecological processes of the region, it becomes possible to identify the phase at which the natural processes have been interrupted. Further, a com-parison of the niche with the corresponding phase of the natural process reveals the significance of the modifications and substitutions of ecological factors. It goes without saying that the determination of the relative importance of the different factors, including the human factor, depends upon the competence of the ecologist.

The time-depth of the niche reveals itself in the equilibrium of different factors and the suppression of modification, with reference to the spontaneous phase. The interventions made within the niche with a view to intensive production presuppose a genuine under-standing of the processes and a consciousness, at the least implicit, of the consequences of over-exploitation. The degree of understanding of the environment may be established by comparing the ecological system and the classification established by the specific terminology of the vernacular language. In the Libinza case, the analysis of the dis-tinctive components defining the various terms reveals a recognition of the functional relations operative between elements of the environ-ment as they emerge from the ecological processes. Similarly, the seasons are classified in relation to the ecological cycle and not simply with respect to the obvious variations in the environment.

The productive activities must be tailored to the existing ecological processes and must not affect them negatively. Overexploitation makes itself felt progressively through various symptoms and provokes a general social reaction, a diminution in the intensity of exploitation.

In distinguishing the particular characteristics of the Libinza niche from those general throughout the zone, it becomes possible, through the analysis of the ecosystem, to refine our understanding of the methods of production and the opposing tendencies within the group. The specific elements of the Libinza niche are related to the development of fishing grounds from which the proprietary kin group obtains a regular supply, and in which other members of the local group are not permitted to participate except insofar as the abundance of fish permits. This is the basis of the permanent opposition within the local group between the agnatic core, which possesses all the rights and transmits them to their descendants, and the affiliated members and clients who will not hesitate to leave if they find better possibilities elsewhere.

The other elements of the Libinza environment, such as the main course of the river, the banks, the open expanses of water free of grass, are, like the seasonal variations, common to the whole zone. An understanding of the principles which determine the seasonal variations is applicable to all the rivers. Everywhere individual fishing is free for all members of the community in the waters of the mainstream and its extensions, and people can move freely from one community to another. The villages along the length of the river have the same appearance throughout the zone. Everywhere in this river environment and not just among the Libinza, the members of local groups who do not belong to the property-holding core experience the same conditions of existence. Their choice of residence is determined by the possibility of associating themselves with the fishing enterprises and commercial relations of a local group more prosperous than their group of origin. They remain constantly alert to the development of any external opportunity which might ameliorate their condition. The one tendency is motivated by the desire to protect and exploit an acquired advantage, the other by a constant desire to speculate in external opportunities. Both can be attributed to a desire to grow richer.

Today, the great fishing grounds have disappeared, but fishing on the Congo river has become very profitable. Since 1960, people have been freely permitted to establish themselves at fishing centres along the whole length of the Congo. Today, only a small proportion of the Libinza population lives permanently in the villages along the Ngiri river.

Footnotes

1. Stereoscopic examination of aerial photographs clearly reveals the different phases in this process on the floor of the Ngiri basin.
2. These indications are important for fishing, as the ichthyological fauna in the pools varies according to their links to the main course of the river.
3. The variations in the level of the water during the unstable period, often of the order of ten centimetres a day, are important to predict, especially when fish baskets are used. The upper part of the basket has to emerge to allow the fish to breathe on the surface.
4. This correct perception of the ecological process throughout the zone of the great rivers and of their swampy extensions has facilitated migrations since it makes possible an understanding of how to exploit the whole variety of aquatic environments.
5. For reasons explained elsewhere (Van Leynseele in press), the working of these fisheries has been progressively abandoned.
6. Wilverth made his voyage in August 1896, thus at the beginning of the *mpela* season, when the waters are about two metres below the top of the flood.
7. A number of measures protected sub-adult specimens including the gauge of the nets and baskets, the use of certain techniques only during a single season, the categorisation of species which gave a different name to sub-adults and protected them by interdict, and numerous food taboos.

References

Bouillenne, R., J. Moureau and P. Deuze 1955. *Esquisse écologique des faciès forestiers et marécageux des bords du lac Tumba.* Bruxelles: Académie Royale des Sciences d'Outre-Mer.

Evrard, C. 1968. *Recherches écologiques sur le peuplement forestier des sols hydromorphes de la cuvette centrale congolaise.* Kinshasa: Instituté pour l'étude agronomique du Congo.

Germain, R. 1965. *Les biotopes alluvionnaires herbeux et les savannes inter-calaires du Congo équatorial.* Bruxelles: Académie Royale des Sciences d'Outre-Mer.

Goffin, A. 1909. *Les pêcheries et les poissons du Congo.* Bruxelles: Ministère des Colonies.

Perdernera, A. 1971. *Carte physionomique de la région de la Ngiri, levé à partir de photos aeriènnes.* Université Catholique de Louvain: document prepared under the direction of J. Wilmet.

Poll, M. 1957. *Les genres des poissons d'eau douce de l'Afrique.* Bruxelles: Ministère des Colonies.

Van Leynseele, P. 1978. *Les Gens d'Eau du confluent Congo-Ubangi.* Unpublished Doctoral thesis, Rijksuniversiteit Leiden.

—— in press. Les modification des systèmes de production chez des populations ripuaires du Haut-Zaire. *African Economic History.*

Vansina, J. 1973. *The Tio kingdom of the Middle Congo 1880–1892.* London: Oxford University Press.

Wilmet, J. 1974. Deux modes différents d'organisation de l'espace en milieu
   équatorial. *Photo Interprétation* **3**, 15–21.
Wilverth, E. 1896. Au lac Ibanda et à la rivière Ngiri. *La Belgique Coloniale* **48**,
   575.

PERMISSIVE ECOLOGY AND STRUCTURAL CONSERVATISM IN GBAYA SOCIETY

PHILIP BURNHAM

At the level of ethnographic analysis, this paper is an attempt to account for the persistence of a set of certain key features of the social structure of the Gbaya people of Meiganga sub-prefecture in Cameroon. At a more theoretical level, discussion of the Gbaya data raises a number of issues that are central to the current debate on the application of ecological concepts to social anthropology, in particular the concept of adaptation.

A First Glimpse of the Problem

The Gbaya people of east-central Cameroon are the subject of a stereotype among government administrators and more '*évolué*' elements in the Cameroon population as being a backward people who are recalcitrant or oblivious to the advantages of 'modernisation'. Prior to Cameroon national independence in 1960, French colonial administrators were also wont to express such views, particularly with regard to the Gbaya habit of continually shifting their village locations and refusing to live in large, stable settlements where they would be more in touch with the infra-structure of colonial development and administration (see Burnham 1975; also Dugast 1949: 140–41). In first undertaking research among the Gbaya in 1968, my attention was naturally drawn to these stereotyped conceptions, and I was interested in examining the veracity of and basis for these views.[1]

I ultimately came to the conclusion that, stripped of their pro-minent racist and other mythic elements, these stereotypes of the Gbaya do contain a kernel of truth. This is not to claim that Gbaya society is simply static and unchanging or that the Gbaya are not 'market-economy minded'; even a cursory glimpse of the last century and a half of Gbaya history is sufficient to dispel any such simplistic notions (Burnham 1975; Burnham in press a and in press b). Rather, despite the turmoil of Gbaya history, the raiding and trading between the Gbaya and the Fulani of Adamawa and the long colonial experience under both German and French rule, there does appear to be a remarkable quality of persistence in certain Gbaya social forms.

To some extent, this persistence can be understood as the result of the marked structural underdevelopment which affects all of the rural areas of north and east Cameroon. On the other hand, factors such as the poor transport network or other such imperfections in the regional market economy are not sufficient to account for the differential persistence of the numerous ethnic groups which make up the population of this large area. Many groups in the savanna regions of Cameroon have assimilated themselves rapidly to the dominant model of settled Fulani society whereas the Gbaya have remained a clearly bounded and discrete social unit (Burnham 1972; cf. Podlewski 1971). The argument developed in this paper is that the continued existence of the structurally discrete entity which we label 'Gbaya society' within the larger Cameroon nation is fundamentally related to, and to a large measure determined by, the conservative effects of a set of social structural elements organised into a system displaying self-stabilising (negative feedback) properties. In the following pages, I will give a rapid ethnographic sketch of the principal elements that make up this conservative nexus and then go on to discuss, in the final section of the paper, the implications of this case for current theories in ecological anthropology.

The Fluid Structure of Gbaya Residential Groups

The Gbaya people who are the subject of this paper inhabit the Meiganga sub-prefecture in east central Cameroon, an area of lightly wooded 'orchard' savanna on the southeastern margins of the Adamawa plateau. This upland region, most of which lies between 900 and 1,500 metres above sea level, is extensively dissected by narrow forest-lined stream valleys. The area is well watered, receiving an average of 1.68 m of rainfall during the seven month wet season from late March or early April through early November.

The 17,000 km² of the Meiganga sub-prefecture are lightly populated (at an average density of about three to four people per square kilometre) by an assortment of ethnic groups. The most numerous group is the Gbaya (about 34,000 in 1966), followed by the pastoral Fulani or Mbororo (about 8,000), the sedentary Fulani or Fulbe (about 6,600), the Hausa (about 2,000) and a half dozen smaller groups. Most Gbaya inhabit small, essentially mono-ethnic villages ranging in size from about 15 to 300 inhabitants (average size about 150) which line the motor roads of the area in accordance with the demands of the government for village relocation and *regroupement* (Burnham 1975). Meiganga sub-prefecture also contains about twenty larger multi-ethnic towns, with populations ranging from 500 to 1,500, which owe their existence to the development of market places and a system of canton chieftaincies introduced in the colonial period. Meiganga town itself, with a population of 6,000 to 7,000 permanent residents (plus a substantial but indeterminate number of transients), is the centre of sub-prefectural government and the largest town in the area. For the most part, the non-Gbaya residents of the district inhabit either these multi-ethnic towns or, in the case of the pastoral Fulani, live dispersed in the bush in family sized encampments.

Despite the outwardly stable impression conveyed by Gbaya villages today, the Gbaya population is in a state of continual flux. Each Gbaya nuclear family spends at least half of every year living away from the 'permanent' villages in scattered encampments alongside their maize fields, either as a single family or in groups of two or three closely related families. Members of the family make weekly visits to their permanent villages to attend church and/or the periodic market, but at least one member of the family must remain behind in the bush to guard the maize fields from the depredations of wild animals. Viewed over the longer term, there is also substantial residential mobility deriving from the fact that individual Gbaya change their villages of residence every nine years on average, and whole Gbaya villages also tend to shift their locations quite often.

This practice of residential mobility is not motivated by pressure on agricultural land resources or other natural environmental considerations. Given the low population density in the region and the high productivity of manioc, the Gbaya staple crop, which is grown on the abundantly available savanna land, Gbaya could live in the same village permanently if they so desired. Likewise, although the cultivation of maize, the Gbaya's main crop, is limited to fields located in the forested stream valleys, there is sufficient land of this type available as well. Instead, Gbaya residential mobility is linked with the non-corporate structure of Gbaya society generally, the weak mechanisms of dispute settlement available in the essentially acephalous Gbaya political structure, and the conflict between an individual's desire to retain autonomy in the political and economic spheres and the levelling pressures exerted by his residential group. This phenomenon of residential mobility will be commented upon more fully at a later stage in this paper.

The Gbaya of Meiganga sub-prefecture are organised on the basis of patrilineal descent into numerous (i.e. a hundred or more) small exogamous patriclans (*zu duk*). These clans are named but non-corporate social categories, and their memberships are dispersed in a number of different clan-based residential quarters (*ndok fu*)[2] which constitute the most significant level of collective action above the nuclear family. Gbaya *ndok fu* contain about 23 inhabitants on average, or about six adult men and their families. Genealogies are seldom remembered beyond two generations, and corporate lineages are not a feature of Gbaya society. Residence with members of the patriclan is the norm, and full brothers and fathers and sons usually reside together throughout their lives. However, degree of genealogical proximity is not a reliable guide to residential choice at the first cousin level and beyond. In a census of Gbaya residential quarters, 96.1% of the 77 men questioned were conforming to the rule of residence with patriclansmen. On the other hand, 38% of Gbaya *ndok fu* are composed of groups of clansmen who are unable to trace genealogical relations with all members of their residential group, a further indication that clanship rather than the lineage principle is the basis for the rule of residence.

Choice of a wife, which is governed by widely ramifying, ego-centred, Omaha-type marriage prohibitions (Levi-Strauss 1969: xxxvi), is a decision taken by the young man and his immediate family and does not enter into any group-focused preferential marriage system. Gbaya marry at an early age: women at about 15 and men a few years later. Bridewealth payments are relatively low compared with other Cameroon societies (i.e. about 30–40% of a family's average annual cash income), and a young man can earn this amount himself in less than a year if need be. But fathers or elder brothers normally provide the bridewealth for a young man's first marriage, and conflict between elder and junior rarely develops over this issue. Indeed, fathers are constantly urging their teenage sons to get married so that they can begin to contribute fully to *ndok fu* welfare. It is the young men who often drag their feet, valuing their few remaining years of freedom before facing the responsibilities of married life. Polygamy, although aspired to by many, is achieved only by about 20% of the currently married men.

Relations within the *ndok fu* are characterised, on the one hand, by strong norms of generalised reciprocity yet all property, including agricultural land, is held as private property. The sharing of food and money which goes on within the *ndok fu* is conceived of as a giving of gifts and an expression of generosity rather than a sharing of corporately held property. These attitudes are understandable when one remembers that despite these norms of *ndok fu* solidarity, the membership of Gbaya *ndok fu* is subject to frequent shifts due to the high rate of residential mobility. Relations of solidarity among clansmen are contingent on co-residence rather than presumptively perpetual in the corporate sense (Maine 1861: 181); *ndok fu* are thus best described as contingent rather than corporate groups. Should an individual or group of clansmen feel that their interests are better served by moving elsewhere, they can count on being accepted in another *ndok fu* of their clan located in a different village. In this new village, they will live under the same conditions of normative solidarity that they had experienced in their previous *ndok fu* of residence. Alternatively, if they are sufficiently numerous to constitute an *ndok fu* on their own, they can move to a village where their clan is not represented and found a new *ndok fu*.

This fluidity of *ndok fu* membership is based on a variety of factors, only two of which I will refer to at this stage: the weak structures of authority in Gbaya villages and the distinctive Gbaya inheritance practices. Dealing with the first factor, it is the case that 'traditional' Gbaya politics, as they existed in the pre-colonial period, have been severely truncated by the enforced peace of the colonial period. Prior to the colonial takeover, Gbaya political structure consisted of loose and shifting patterns of intra and inter-clan alliances under the leadership of Gbaya war leaders who were in continual competition for followers (Burnham in press b; also Burnham 1975: 577–79). The French did away with this system and replaced it with appointed village headmen, who even today exercise little control over the members of their village since they have no effective sanctions at their disposal. The rate of turnover in the village headman role is consequently extremely high, and headmen are unable to prevent frequent secessions from their villages.

The Gbaya inheritance system likewise contributes to withdrawals from the village, since it has the effect of distributing a man's estate widely while favouring senior agnates at the expense of juniors.

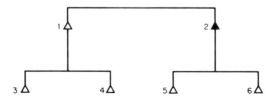

Fig. 1. Order of Gbaya inheritance.

Estates are shared among all adult co-resident clansmen according to principles of generation and seniority within a generation. Estates are divided into 'naturally occurring' divisions — a cow, a maize field, or a radio, and the shares are distributed in turn to the dead man's agnates according to the order shown in Figure 1. If genealogical relations are not traceable among all co-resident clansmen, an artificial genealogy is created which assimilates the unconnected segment according to which segment is considered to be the founder of the *ndok fu* and according to the chronological ages of the senior men in both segments. The Gbaya system of inheritance is a complex phenomenon which I can only briefly sketch here, but its implications are plain from even this short description. The tendency toward residential mobility is enhanced by the fact, apparent in the figure, that junior segments tend to lose out in the inheritance distribution since the best shares go to the senior segment. Thus, for example, if man 2 in the diagram dies, his own sons (5 and 6) must wait until agnates 1, 3 and 4 have had their pick of the shares before they receive their portions of the estate. An estate is never liquidated and the cash shared out equally and, as a result, the 'shares' of an estate tend to be of decreasing size as the distribution proceeds.

For these reasons, as well as for others to be specified later in this paper, Gbaya collective behaviour above the family level (or set of full brothers and their families, when present) can best be described as fluid or contingent in structure. That is to say, rather than being characterised by widespread patterns of presumptively perpetual relations between individuals and groups, Gbaya society is built on the presumption that most social relations are predictably impermanent and short-lived. Only the clans are presumptively perpetual, and these are only categories from which a number of impermanent residential quarters (*ndok fu*) may be recruited and are not corporate groups themselves. Despite the publicly proclaimed solidarity of *ndok fu* groupings, therefore, Gbaya social relations display an underlying individualism which is expressed in highly individuated property relations and weak authority structures. Even at the nuclear family level, as mentioned earlier, there is an easy multiplication of family units as men are encouraged to marry and establish their own families at an early age. As we shall see in a moment, such newly created families have few problems in setting up an independently viable domestic economy, due to the permissive Gbaya subsistence economy.

A Permissive Subsistence Economy

Gbaya experience little difficulty in assuring themselves an adequate subsistence. Although manioc, their main food, has acquired a poor reputation among nutritionists as a staple crop, this judgement is not borne out with regard to the role of manioc in Gbaya nutrition. Admittedly, however, the use of manioc in the Gbaya economy con-stitutes a rather special case in this regard. Manioc is a very productive crop, and Gbaya farmers produce about fifteen tonnes of manioc root per hectare on average (after 14 months growing time). Even in the poorest conditions, yields of five tonnes per hectare are attained. Manioc crop failure is virtually unknown in the present day.

Vacant land for manioc cultivation is very plentiful and is free for the taking by anyone, being subject to no collective or residual rights of tenure. Moreover, Gbaya techniques of manioc cultivation require only limited labour since the savanna grasses are simply burnt and scraped away and the soil is not turned over by hoeing. Manioc can be planted at almost any time during the seven month wet season, and the occasional weeding of manioc fields can also be fitted in as convenient among more pressing seasonal chores. The limit on Gbaya manioc production derives from the limited demand for the crop in the cash marketing sector rather than from any constraints internal to the production process itself.

In these conditions and given the extremely low population density of Meiganga sub-prefecture, there is no pressure on the carrying capacity of agricultural land for the manioc crop.[3] Manioc accounts for the major part of Gbaya daily calorie intake the year round and since the roots can be left in the ground until needed, there is no question of seasonal hunger periods.[4] Manioc leaves also provide an important source of vitamins when eaten in sauces. But the other main contribution of manioc in the Gbaya subsistence economy is that it forms the basis for a 'symbiotic' exchange of starches for meat and other animal products between the Gbaya cultivators and the many Fulani herders living in the district. These exchanges occur both directly in kind in the context of bond friendship relations between Gbaya and pastoral Fulani herders and indirectly, via the medium of cash, in market place sales of manioc and beef.

Primarily because of this highly institutionalised inter-ethnic exchange, the Gbaya eat substantial amounts of meat, and any possibility of the protein deficiency diseases that are sometimes associated with manioc-eating populations is dispelled.[5] The Gbaya diet is also supplemented by a wide variety of cultivated and gathered secondary foodstuffs, the Gbaya's continued exploitation of bush products even in the present day being quite noteworthy. A few minor elements of day-to-day consumption are provided from a small but regular cash income derived from the production and sale of minor craft items such as mats, basketry, pottery, rope and wooden utensils. These activities do not offer the possibility for major capital accumulation but do provide the pocket money for purchasing beer, tobacco, salt and other sundries.

In summary terms, therefore, I describe the Gbaya subsistence economy as 'permissive' in that the activities involved in ensuring an adequate and predictable food supply entail relatively few constraints or specifications for the social behaviour of Gbaya individuals. There is no scarcity of the necessary factors of production. No corporate structures of landholding or labour organisation are employed. There is little rigidity in the scheduling of subsistence related activities. And, as a result, individual nuclear families, established when the partners are still quite young, can immediately be self sufficient in subsistence if they so desire.

Opportunities and Constraints in the Market Economy

In addition to manioc cultivation, which is principally oriented toward the production of a family's subsistence via direct consumption and exchanges for beef and other food items, the other main Gbaya crop is maize. Only a small proportion of the Gbaya maize crop is consumed by the Gbaya themselves, primarily in the form of beer, and the major part is sold in bulk to non-Gbaya transporters who export the crop from the Meiganga sub-prefecture. Unlike Gbaya manioc cultivation where relatively weak demand imposes a limitation on the size of plots cultivated, demand for maize is strong, and Gbaya farmers attempt to maximise their production. The principal constraints on their efforts are the large labour requirements for clearing maize fields in the forested stream valleys and the inadequate supply of labour at this phase of the agricultural cycle. Even if a Gbaya has cash available to hire extra help, there is little assurance that he will be able to find hired help. Land in the forested stream valleys which is used for maize cultivation is still in plentiful supply, except in the vicinity of a few of the larger towns. The Gbaya practice of moving to live alongside their maize fields during the farming season means that stream valleys in quite a wide radius from a village are available for use. Once cleared of primary forest, a maize field becomes the personal possession of the farmer and may be rented sold, or inherited as freehold property. As in the case of manioc land, there are no residual rights exercised by the collectivity over maize land.

Gbaya maize cultivation is subject to a variety of risks which may reduce or totally destroy the harvest. These risks include incomplete burning of the felled forest and attacks by animal pests such as termites, guinea fowl, and monkeys. So substantial and unpredictable are these risks that about 29% of Gbaya maize farmers completely fail in their efforts in any one year while a fortunate 5 to 10% profit greatly, earning as much or more from their maize crop alone as an average Gbaya family is able to earn from all its income earning activities in one year combined. Such success is very unpredictable, however, and a maize farmer's harvest in one year is not a reliable indicator of the outcome in the subsequent year.

Aside from maize farming, other sources of cash income available to Gbaya include honey gathering, cattle investment and butchery, wage labour, and petty trading. Honey gathering, involving the construction of artificial hives as well as the exploitation of naturally occurring bee trees, is undertaken by many men in a small way and yields minor amounts of income. Extensive honey gathering, which involves large amounts of labour preparing many hives and a great deal of difficult and dangerous tree climbing at night accompanied by numerous stings, can yield large profits of the same order of magnitude as those earned by the more successful maize farmers. However, the labour requirements for a major effort at honey production conflict to a considerable degree with those involved in maize production, and the two activities therefore constitute alternatives (of which maize farming is clearly preferred). Few Gbaya men engage in large scale honey gathering and then only for a year or two to pay a brideprice or achieve a similar savings target.

Cattle investment and market trading, although practised in minor ways by many Gbaya, do not yield large profits unless a man has a substantial sum to invest. In other words, in both economic activities, there is a threshold of capital accumulation which must be reached before these enterprises can contribute in any major way to Gbaya incomes. For a number of reasons, related both to the highly variable degree of success experienced by the Gbaya in their main cash earning activity of maize farming and to the effects of certain wealth levelling mechanisms that operate within Gbaya residential groups (which will be discussed more fully in a moment), Gbaya find it difficult to save substantial sums, and Gbaya penetration of these income earning activities is consequently quite limited.

Opportunities for wage labour are not common within the Gbaya economy. Gbaya sometimes hire other Gbaya as seasonal wage labourers in connection with maize growing, but most Gbaya farmers have little cash available to hire labour during the early part of the maize growing cycle when labour requirements are greatest. Then too, able bodied Gbaya usually prefer to cultivate their own maize fields rather than to work for others. At the times of the year when there is a surplus of labour in the Gbaya economy, the main possibility for employment is to be found with Fulani employers, who hire Gbaya to build houses or to cultivate millet. Gbaya men sometimes under- take such employment for short periods if they are badly in need of cash but considerable stigma attaches a working for Fulani since, in the past, they had used Gbaya slaves for such tasks. Indeed, so reticent are Gbaya to engage in this employment that many of these jobs are filled by non-Gbaya labour migrants from the neighbouring Central African Empire who are attracted to Meiganga sub-prefecture particularly for this purpose.

Summing up the economic opportunities available to the Gbaya, one can say that it is easy for a Gbaya family to establish itself as an independently viable economic unit but it is very difficult for them to accumulate a significant amount of wealth. A small but regular income is available from sales of manioc and craft products, but this serves only to ensure an adequate nutrition and purchase house- keeping necessities. More substantial lump-sum earnings are possible in maize farming, an economic opportunity equally available to all, but the unreliability of this income leads Gbaya farmers to treat such profits more as windfalls than as the basis for a permanent upward adjustment of their standard of living. Occupations such as full-time market trading and large-scale cattle husbandry are not feasible for the great majority of Gbaya because of the high capital investment required and are also defined as being outside the traditional range of Gbaya competences.

Gbaya attempts at wealth accumulation are blocked not only by the technical chracteristics of Gbaya production activities but also by the wealth dispersal mechanisms inherent in Gbaya residential group structures which have been referred to earlier. As contingent rather than corporate groups, Gbaya *ndok fu* villages do not operate on a presumption of perpetuity but rather employ canons of generalised intragroup reciprocity as tests of individual members' continued commitment to the group. Although all property is individually owned in Gbaya society, Gbaya norms prescribe a wide- spread sharing of such property within an individual's co-resident group. This principle is most clearly manifest in the practice of communal eating within *ndok fu*, in which the wives of the male members prepare collective meals for the men in a loose rotation. In the modern context, a man's cash earnings are not exempt from such demands from co-resident clansmen, and there is a continual flow of cash gifts, loans, and mutual aid within Gbaya residential units. These practices do provoke considerable ambivalence in the minds of the Gbaya participants. Demands for sharing, particularly of cash and capital goods, stand in direct conflict with individual efforts to accumulate wealth and penetrate new spheres of economic opportunity, and Gbaya often practise deception in their attempt to retain their earnings for their own use.

The practice of generalised reciprocity as a measure of continued commitment to one's residential group is one which is hedged about by both supernatural and practical sanctions. Insufficient generosity to one's co-residential agnates can lead to the offender being subject to a curse and mean behaviour is also taken to be symptomatic of witchcraft. These strains are particularly noticeable when *ndok fu* grow larger and generalised reciprocity becomes correspondingly more onerous. Any reduction in involvement in the network of reciprocity is, in turn, interpreted as signalling a reduced commitment to the residential group and the imminent probability of its dissolution. With residential mobility an easy option in Gbaya society, Gbaya actors need make little effort to smooth over emergent strains and conflicts and have few compunctions in shifting residence if it seems advantageous to do so.

The Conservative Nexus of the Gbaya System

Although the preceding pages have provided only a brief ethnographic sketch of Gbaya social organisation, the data should be sufficient for the reader to discern the interlinkage of structural features which generates the conservative tendencies and the ethnic boundedness of the Gbaya system. This set of interlinked structures includes the individuated property and labour relations, the non-corporate, non-centralised political structure, the non-gerontocratic marriage structure and the consequent early age of marriage and easy multiplication of family units, which together ensure quite a high degree of autonomy for men and their families. Combined with a permissive ecology, in particular a low population density in relation to the productivity of Gbaya agriculture, these structures permit and encourage a high rate of residential mobility for nuclear families or sets of full brothers and their families which, in turn, reasserts the tendencies toward individual autonomy within the system. The loose structure of clan categories and the rule of residence based on clanship provides a supple but unchanging normative framework within which this continual process of individual reassortment can take place. Thus, while the structural framework remains unchanged, at the inter-personal level the patterns of social relationship are contingent and unstable (or what I have termed 'fluid').

Although to some extent, the phenomenon of easy residential mobility has been singled out for particular attention as a regulatory mechanism in this conservative structural nexus, it is in the nature of circular causal (negative feedback) relations that no one element can be singled out as the sole determinant of the system's self-persisting effects. In fact, mobility is simply the main 'triggering mechanism' which acts to relieve pressures on an individual's autonomy, pressures which are continually being generated by a variety of social processes both internal and external to the conservative nexus.[6] Not only do Gbaya use residential mobility as a means of escape from externally generated administrative pressures exerted by the Cameroon government authorities (as they did from the Cameroon government's colonial and pre-colonial predecessors, the French, Germans and Fulani) but they also employ this mechanism to escape from local level disputes and from the progressively more onerous demands for sharing of material wealth which expand in step with the demographic growth of residential units. These pressures on the individual have their cultural as well as their material dimension and include the dangers of witchcraft and other supernatural attacks which can often be evaded by shifting residence.

The norm of reciprocity which obtains among co-residents, particularly co-resident clansmen, is linked in with the conservative nexus in several ways. On the one hand, it acts as a wealth-levelling mechanism which severely limits the capacity of Gbaya actors to engage in the more lucrative occupations available in Meiganga sub-prefecture, which require substantial capital inputs. As noted above, Gbaya actors respond negatively to such pressures, eventually moving to join a new residential group when their old residential group has grown too large and the demands for sharing become too oppressive. But joining a new residential group does not end such demands altogether; it only lessens them. And Gbaya are faced with the fact that no matter in what Gbaya village they live, they must always participate in an intensive network of generalised reciprocity and thus they will always be subject to these wealth-levelling effects. Escape is possible only by moving out of the rural Gbaya village setting into an urban area or multi-ethnic town where demands for wealth sharing can be more successfully evaded without incurring traditional forms of social control. But the decision to opt out in this way has clear implications for an actor's ethnic affiliation and in this sense, the Gbaya norm of generalised reciprocity acts as the principal ethnic bounding mechanism which continues to generate a large measure of structural independence for the 'traditional' Gbaya social system vis-à-vis the wider regional, national, and international systems. To sum up the argument in one sentence, the structural features listed above constitute a self-perpetuating, self-bounding system which acts to block tendencies toward differentiation in wealth or development of new corporate or other collective structures in rural Gbaya communities.

Social Ecology and Self-Regulating Systems

In the preceding section, I have sketched out the main structural features in Gbaya society which I consider to be the sources of its self-persisting character. These structures do not exist simply as a collection of unrelated parts but mutually interact in a systemic manner so as to constitute a nexus of structural persistence within the varied flow of Gbaya history. To the extent that this system is based on an interlinkage of both environmental and social structural variables, it is ecological in character, yet it remains to be seen whether the use of the label 'ecological' in this context should be taken to convey anything more than this essentially descriptive characterisation.

For certain ecologically minded anthropologists, in particular those who have sometimes been grouped under the label 'neo-functionalist', the study of self-regulating systems has become the central focus of study. In attempting to avoid the classic pitfalls of functionalism while still employing the concept of 'adaptation' as their principal analytic tool, authors such as Rappaport, Vayda, and others have bent over backwards to adhere to the strictures of Hempel (1958), Brown (1963), and the other philosophers of science (cf. Merton 1949: 51–55) who have specified a highly rigid set of procedures for the framing of acceptable forms of functional explanation.

A recent example of this school of thought is provided by Rappaport's response to Friedman's (1974) critique of the neo-functionalist position. In his article, Rappaport (1977: 142, 148–49, 167–68) explicitly identifies the discipline of ecological anthropology as the study of self-persisting systems. Recognising that self-persistence in a system implies nothing about its adaptiveness (if 'adaptiveness' is defined in its traditional Darwinian sense), Rappaport takes the unfortunate step, in effect, of redefining adaptation to fit his own purposes. Adaptation, now defined much more generally as any tendency toward self-maintenance within specifiable limits, is distinguished from homeostasis, which is permitted to keep its normal biological meaning of self-maintenance within the limits of viability of the system.[7] In twisting and turning to avoid Friedman's verbal shafts, Rappaport not only produces a heuristically oversimplified scheme of analysis which is clearly incapable of dealing with social change or any other social phenomenon that does not fit into his neat negative feedback model but also seems to forget that anthropology's major goal is to explain social behaviour. As Friedman noted in his 1974 critique (and as I have also argued — Burnham 1973: 98, 99), analyses such as Rappaport's (1968) of the Maring, while claiming that social behaviour is controlled and selected for by techno-environmental limits (see Rappaport 1977: 140), are unable to demonstrate that the environmental limits in question are connected in any meaningful way (i.e. are processurally linked) with the social behaviour in question. Instead, in the face of Friedman's eminently plausible argument that the Maring pig population is regulated by social rather than environmentally conditioned factors (Friedman 1974: 459–60), Rappaport has no real rebuttal other than to assert that social limits and environmental limits can be expected to coincide. Thus Rappaport (1977: 161) states.

> In small undifferentiated societies such as that of the Maring, in which all adult members are directly and continuously engaged with the natural environment and in which corrective informational feedback from the environment is direct, observable and relatively quick, we may expect a close correspondence between culturally specified motives and goals and the requirements of ecosystemic processes.

In short, Rappaport is simply telling us that, at least among primitive societies, we can trust in the ecological wisdom of the actors. This is a familiar old theme in the ecological literature (Burnham 1973: 95), and its reassertion by Rappaport is one further indication that, in practice, Friedman (1974: 457) was essentially correct when he claimed that 'the new functionalism is fundamentally the same as the old functionalism'.

Conservatism and Change in Gbaya Social Ecology

In using the concept of negative feedback to describe the inter-relationship and persistence of a set of variables in the Gbaya social system, therefore, I am making no statement as to the adaptive value of this structural nexus. Indeed, as I have made clear already, I consider application of the term 'adaptation' to social systems to be premature at best and generally quite misleading. For me, negative feedback refers simply to a particular pattern of interaction of variables in a system which contributes to the self-maintenance of these variables.[8] Along with Friedman (this volume), I would reject Rappaport's application of the concept of control hierarchy and his contention (Rappaport 1977: 172) that

> Unified adaptive systems must regulate very large numbers of variables through cybernetic and other mechanisms, but no adaptive system could possibly be a mere heap of regulators regulating the states of variables through the initiation and termination of response sequences. Special adaptations must be related to each other in structural ways, and adaptive systems must take the form of complex sets of regulatory operations, hierarchically organised, with regulating regulators and others regulating them. Adaptive systems are not only cybernetic but hierarchical.

Recent contributions in the field of stability theory (Lewontin 1969; Holling 1973) have argued that the existence (within limits) of a stable relationship between variables in a complex system, such as that displayed in the Gbaya case, is not necessarily dependent on cybernetic programming or control hierarchies.

Neither have I argued for any tendencies toward eco-systemic regulation in the Gbaya system with respect to environmental tolerance limits. On the contrary, if present trends of demographic growth are projected into the future, the carrying capacity of the Gbaya agricultural system will eventually be exceeded. In general, it seems highly unlikely that a society such as the Gbaya, which is based on a presumption of geographical mobility and expansion in the context of low population density, contains within itself self-limiting mechanisms which would effectively regulate populations as the carrying capacity limit is approached. Whatever the exact nature of the social changes induced by such land pressure, I would predict that they would qualitatively alter the fluid and permissive character of Gbaya social ecology.

At present, no such fundamental restructuring of the Gbaya system is in evidence. However, this does not mean that Gbaya society is entirely static[9] or that the negative feedback system just described is a complete account of the processes going on in Gbaya society. As I have already mentioned, Gbaya history is complex and in many respects, including conversion to Islam or Christianity, the development of a system of market places and cash cropping, the institution of canton chieftaincies, the introduction of western education, etc., Gbaya society has changed and continues to change over the past century. Nonetheless, during this time, the relationship of the variables which constitute what I have termed 'the conservative nexus' of the Gbaya system has not changed. However, as we have seen, the social boundary established by the conservative nexus is simply one of forces of social control and conformity acting on rural Gbaya village. Individual Gbaya can cross this boundary at any time if they so desire and at present, the main social change phenomenon in Gbaya society consists of a movement of actors over the structural boundary separating the conservative nexus from the opportunities available in the wider Cameroon national system. In most cases, this 'movement over the boundary' is both a spatial as well as a structural movement, as individuals change their places of residence from the limiting confines of rural agricultural villages to the more economically and socially varied contexts of town or city life. The rate of movement over this boundary is determined by a large number of factors including the degree of success of individual Gbaya farmers in producing and marketing their harvest and constituting a sufficiently large capital base for investment, the availability of salaried occupations, educational opportunities and other external inputs such as pensions and remittances of savings from urban Gbaya, as well as the degree of political and economic advantage accruing to Fulbe as compared to Gbaya ethnic identity (see Burnham 1972). At the date of my field research, this rate of movement was relatively slow, since few Gbaya farmers were successful in clearing the economic threshold and the Cameroon government did little to stimulate local economic opportunities. Barring radical external intervention in Meiganga and given the low population density of the region, it seems probable that the present Gbaya social ecological adjustment will continue to provide a stable way of life for its adherents for a good many years to come.

Footnotes

1. My field research among the Gbaya was carried out during twenty-one months between 1968 and 1970 and again for short periods totalling four months in 1973 and 1974. Financial support for this research was provided by the University of California at Los Angeles, the Wenner-Gren Foundation, and the Hayter Fund of University College London.
2. The '*o*' in the Gbaya term *ndok fu* is an 'open o' and is pronounced like 'awe'.
3. Although the calculation of the carrying capacity of the Gbaya agricultural system is subject to certain methodological problems (cf. Brush 1975), there can be little doubt that Gbaya population is well below capacity. The population density averages only about three to four persons per square kilometre in the Meiganga subprefecture and even at minimal yield levels, one hectare of manioc provides an output sufficient for the nutritional and marketing requirements of 20.6 people for one year, at average levels of usage.

A more detailed discussion of Gbaya carrying capacity is beyond the scope of this paper.

4. Nutritional data on the Gbaya of Meiganga are reported in Winter (1966), although the FAO minimal daily nutritional requirements used in Winter's calculations are probably too high (cf. Waterlow & Payne 1975 and Winter 1966: 72).
5. A clinical study of the nutritional status of the Gbaya and other peoples of the Department of Adamaoua in Cameroon detected no cases of kwashiorkor and only occasional signs of undernourishment (Winter 1966: 70–71).
6. See Stauder's discussions (1971 and 1972) of Majangir society, where residential mobility plays a similar role in the context of a permissive ecology.
7. I believe that this distinction between adaptation and homeostasis is clearly implied in Rappaport's recent discussion (1977: 162–66) of formal versus final causality, in his recognition (1977: 167–69) of a broad class of systems with 'adaptive' properties, ranging from 'simple animals' to 'elaborate empires', as well as in his statement that 'negative feedback need not be homeostatic' (1977: 159). On the other hand, in places he still conflates the two concepts (Rappaport 1977: 168–69) and thus continues to fall prey to the biologism about which I have previously warned (Burnham 1973).
8. In a discussion of this paper with Jonathan Friedman, he regretted my continued use of the term 'negative feedback' in view of its close association with notions of cybernetic self-regulation. While I clearly agree with the spirit of his argument, I am not sure that one must go so far as to avoid this otherwise useful concept altogether.
9. In employing the term 'stability' (or 'conservatism' as its synonym) in this paper, I am using it in the sense employed by Lewontin (1969: 19–21, see especially his discussion of local and relative stability).

References

Brown, Robert 1963. *Explanation in social science.* Chicago: Aldine.

Brush, Stephen 1975. The concept of carrying capacity for systems of shifting cultivation. *Am. Anthrop.* 77, 799–811.

Burnham, Philip 1972. Racial classification and identity in the Meiganga region: North Cameroon. In *Race and social difference*, P. Baxter and B. Sansom (eds.). Harmondsworth: Penguin Books.

—— 1973. The explanatory value of the concept of adaptation in studies of culture change. In *The explanation of culture change*, C. Renfrew (ed.). London: Gerald Duckworth.

—— 1975. *Regroupment* and mobile societies: two Cameroon cases. *J. Afr. hist.* 16, 577–94.

—— in press a. Notes on Gbaya history. *Contribution de la recherche ethnologique à l'histoire des civilisations du Cameroun*, C. Tardits (ed.). Paris: C.N.R.S.

—— in press b. Raiders and traders in Adamawa. *Asian and African systems of slavery*, J. Watson (ed.). Oxford: Basil Blackwell & Mott.

Dugast, Idelette 1949. *Inventaire ethnique du Sud-Cameroun.* Douala: I.F.A.N.

Friedman, Jonathan 1974. Marxism, structuralism and vulgar materialism. *Man (N.S.)* 9, 444–69.

Hempel, Carl 1958. The logic of functional analysis. In *Symposium on sociological theory*, L. Gross (ed.). Evanston: Row, Peterson.

Holling, C.S. 1973. Resilience and stability of ecological systems. *Annual Review of Ecology and Systematics* 4, 1–23.

Lévi-Strauss, Claude 1969. *The elementary structures of kinship*. London: Eyre & Spottiswoode.

Lewontin, Richard 1969. The meaning of stability. In *Diversity and stability in ecological systems*. Brookhaven Symposia in Biology No. 22, 13–24.

Maine, Henry 1861. *Ancient Law*. Boston: Beacon.

Merton, Robert 1949. *Social theory and social structure*. Glencoe: Free Press.

Podlewski, A.M. 1971. La dynamique des principales populations du Nord-Cameroun. *Cahiers O.R.S.T.O.M.*, sér. Sciences Humaines, 8.

Rappaport, Roy 1968. *Pigs for the ancestors*. New Haven: Yale University Press.

—— 1977. Ecology, adaptation and the ills of functionalism. *Michigan Discussions in Anthrop.* 2, 138–190.

Stauder, Jack 1971. *The Majangir*. Cambridge: University Press.

—— 1972. Anarchy and ecology: political society among the Majangir. *S. West. J. Anthrop.* 28, 153–68.

Waterlow, J.C. and P.R. Payne 1975. The protein gap. *Nature* 258, 113–17.

Winter, Gérard 1966. *Le niveau de vie des populations de l'Adamaoua*. Yaounde: O.R.S.T.O.M.

PASTORAL PRODUCTION, TERRITORIAL ORGANISATION AND KINSHIP IN SEGMENTARY LINEAGE SOCIETIES

PIERRE BONTE*

Since utilisation of grassland environments by pastoral or semi-pastoral peoples represents one of the more specialised human ecosystems, the study of such societies provides an excellent opportunity for analysis of the correspondence between environmental systems and social systems. Typically, these pastoral ecosystems are based on the exploitation of the annual or perennial plant cover (the primary producers) by the herds of herbivorous animals (the primary consumers), with man playing the role of secondary consumer in competition with different predators. On this ecological system, a variety of different forms of subsistence and social organisation can develop and in this article it is not my aim to seek for a one-to-one correspondence between types of ecosystem and types of social system. Rather, my aim is to discuss, with reference to a series of examples, the significance of ecological data and theories for anthropology and to touch on a number of other problems which will emerge in the course of this discussion.

At the risk of oversimplification, one can say that the essential characteristic of pastoral exploitation of these grassland environments is the intensive application of labour to the herds (domestication). On the other hand, human efforts at amelioration of pasturage are much less intense. They are not, however, exclusively predatory, since the preparation of watering points, selective burning, selection of pasturage, etc. also represent a certain degree of domestication of the vegetation. Because of these factors, pastoral nomadic societies, and to a certain extent other societies in which production is predominantly pastoral, must resolve the problem of dealing with two different levels of appropriation of nature. The appropriation of the herds takes place at the level of the domestic group or even the individual; the appropriation of pasturage takes place at a collective

---

*Translated from the French by Philip Burnham.

level in the context of what I shall henceforth term the pastoral community (Bonte 1973; 1978a).

It was also with reference to a predominantly pastoral society (the Nuer) and a fully pastoral society (the Bedouin of Cyrenaica) that E.E. Evans-Pritchard presented his analysis of segmentary lineage societies. This analysis has subsequently developed into a general theory concerning forms of social organisation in which kinship is dominant. This theory can be summed up in several propositions:

1. Of the various forms of kinship relation, descent is of particular structural significance and fulfils a variety of functions.
2. Unilineal descent defines the form of individual social groups (lineages) and furnishes a model for the totality of social relations, e.g. relations between territorial groups.
3. Lineage segments and territorial groups are genealogically defined according to the rules of complementary opposition (i.e. competition at one level and association at a higher level).
4. These rules of complementary opposition form the basis for the entire structure of the society which is expressed in political terms as a balance of power between segments.

Segmentary lineage theory has progressively emerged as a general sociological model for the whole category of 'stateless' societies whose tendencies toward internal disequilibrium are continually counterbalanced by the fission and fusion of their segments.

Segmentary lineage organisation is able to cope with the necessities of pastoral production. It permits, in particular, the maintenance of the autonomy of the families who control the herds while at the same time assuring their integration into a larger community structure which guards the rights of access of each of the families to collectively utilised resources. The families constitute elementary segments in the segmentary genealogical structure. This structure, taken as a whole, defines the system of rights over an estate, including particularly pasturage, to which each family may lay claim by reason of its genealogical position. The number of pastoral and semi-pastoral societies which display such a segmentary lineage structure is indeed large: aside from the Nuer and Bedouin studied by Evans-Pritchard, one can cite the pastoral peoples of the Maghreb, those of Iran and Afganistan and, more generally, the majority of the pastoral societies of the circum-Mediterranean.

As I have already pointed out, my aim in this paper is not to search for interrelations between a particular ecological system and a particular form of social organisation. Rather, I hope to analyse the manner in which anthropology has made use of models and explanations drawn from ecology. In this regard, the development and elaboration of segmentary lineage theory is quite revealing. As I have tried to show in a previous article (Bonte 1978b), this theory was only constructed at the price of a considerable simplification and selectivity of the data. From an epistemological point of view, these difficulties in the handling of the empirical data are linked with an ambiguity in the theory itself. Are we dealing with a specific model designed to aid us in understanding particular societies at particular points in time (to a large extent this is the case in the study of the Nuer) or is it a question of a more general model of the functioning of 'stateless' societies? Within the theory, the role of ecological factors is marked by the same ambiguity. On the one hand, the ecological variables are some of a number of determinative factors which together constitute a model especially designed to analyse Nuer society. On the other hand, these ecological variables also account for some of the variations within the general segmentary lineage model. With particular ecological conditions leading to the adjustment of the general model, the model itself may be defined on the basis of a collection of locally adapted forms. It is therefore apparent that, through this concept of adaptation, a notion of determination in the last instance by the environment has been introduced. The adaptive functioning of the model is thought to bear witness to its correspondence with reality and its effectiveness: the system functions because it is adapted. We see here a new variant of the functionalism which has dominated anthropology for so long. By this means as well, we see concepts such as equilibrium being introduced, or rather, reintroduced.

Rather than continue this abstract discussion, I prefer to illustrate and extend these introductory remarks by reconsidering the debate from *Man* concerning the origin of the Nuer and the Dinka and the differences between their social systems. This debate has value in the present circumstances not only because it permits us to reexamine the data on one of the semi-pastoral societies which is at the origin of segmentary lineage theory but also because it is very closely centred on the problem of the significance of ecological determination and will permit us to consider further some of the epistemological problems already mentioned.

Nuer and Dinka: Variations on a Segmentary Theme

Are the Nuer Dinka? If so, what are the causes of the differentiation of the two populations? These are the initial terms of the *Man* debate.

In addition to these questions, the *Man* exchange raised once again the problem initially posed by Sahlins in his much quoted but little utilised 1961 article: what is the degree of generality of the segmentary lineage model? The date of publication of Sahlins' article is not without significance. It appeared at the same time as Leach's (1961) criticism of the overemphasis in the anthropological literature on the analysis of unilineal descent groups, a topic which at that time had reached its apogee but which was also touching its limits. In his article, Sahlins underlined the specialised characteristics and limited distribution of segmentary lineages. In his view, segmentary lineages were linked with a particular form of intertribal relationship: the conquest of one group by another. The organisation of unilineal descent groups in a relation of complementary opposition was favourable for the political mobilisation of the conquering group and would confer upon it a military superiority. But conquest would eventuate in the disappearance or the absorption of the conquered populations, unless it resulted in the constitution of a more developed form of social organisation based on a chiefdom. In either case, conquest would create the conditions for the disappearance of the segmentary lineage organisation. In addition to the Tiv example, the case of the Nuer and their conquest of the Dinka was one of the examples chosen by Sahlins to illustrate his argument.

Newcomer's (1972) article extended Sahlins' argument but also inverted his perspective to some extent, Newcomer, like Sahlins, argued that their segmentary lineage organisation gave the Nuer a superiority over their neighbours, but he rejected any idea of a conquest by the former of the latter. In fact, according to Newcomer, Nuer and Dinka have a common origin. The Nuer are simply Dinka who have experienced a profound social mutation due to the development of this new form of organisation. Newcomer hypothesised that the origin of this segmentary lineage organisation was due to the occupation of a less favourable habitat by groups of the Dinka type. Occupation of this more difficult environment necessitated wider cooperation and more effective political alliances.

The principal contribution of Newcomer's short article to this debate was to get rid of the fixed concept of social groups and systems constituted *ad eternam* and to link social group formation and ethnic differentiation to a social dynamic, in particular to changes in material conditions such as the ecology (Newcomer 1973). The article, however, had its evident limitations. Although it dispensed with certain of Sahlins' historicist hypotheses (such as conquest), it continued to explain the appearance of segmentary lineage organisation on the basis of the superior adaptive qualities of this form of organisation. Newcomer's recourse to a biological vocabulary underlines this fact. Although Newcomer (1972: 11) rejects any notion of environmental or technological determinism, ecological factors intervene from the outside to justify change toward those 'social mutations' which are best adapted. In some respects, this argument is reminiscent of social Darwinist thinking.

Sahlins' explicit and Newcomer's implicit evolutionist biases, which assert the superiority of the segmentary lineage as a form of organisation, were criticised by Glickman in his 1972 article. This article also underlines the importance of the unique ecology of Nuerland. Because of the considerable distances that separate wet season from dry season pastures and the periodic and variable flooding of the country by the two branches of the Nile, patterns of trans-humance and inter-group cooperation must be organised on a wider scale. Yet this does not necessarily imply a 'superior' adaptation. Moreover, Glickman calls into question the notion of a conquest of the Dinka by the Nuer, if conquest is to be taken to mean a more or less well organised occupation of territory. Territorial expansion is rather the eventual outcome of complex processes of incorporation of Dinka at the Nuer periphery. This absorption occurs in the context of the continual movement of both groups and individuals within Nuer society — movements which are mainly linked with cattle. There is no particular institutional cause of the military superiority of the Nuer but rather a warlike tendency which is also related to the role of cattle, the goal of Nuer military expeditions.

Consequently, it is no more the political and military superiority of the Nuer than their desire for territorial expansion which can explain their heavy pressure on the Dinka. For here again we must look at the data carefully. Even within Nuer society itself, segmentary lineages only really developed among the eastern Nuer who occupied this zone in the 19th century and who continue their territorial expansion right up to the present. Segmentary lineages are less well developed among the western Nuer, whose social system is closer to that of the Dinka. The idea of a direct determination by ecological factors is thus doubly called into question. Neither does the segmentary lineage appear in the area which best displays the ecological constraints which are supposed to produce it nor can the segmentary lineage be said to constitute a precondition for territorial expansion since we only observe it among the groups which have most recently settled in their territories.

We must therefore find some other form of explanation in order to understand both the influence of ecological factors on segmentary lineage organisations and the relation between the conditions giving rise to segmentary lineages and the process of territorial expansion. First of all, I should like to return to a consideration of the nature of the ecology of Nuerland. Are the ecological conditions in Nuerland today truly so difficult? Southall (1976) has noted that this marshy region, which is subject to periodic inundations from the different branches of the Nile, constitutes a difficult but nonetheless favourable human habitat for the development of pastoral production. The Nuer themselves are quite conscious of this fact. The environment has another important characteristic: variations in the level of flooding lead to variations in available pasture. Consequently, the animal and human populations can increase rapidly, but the cattle population can also undergo severe declines. This phenomenon is not without effect on the humans themselves. I would advance the hypothesis that the relation which exists between the development of segmentary lineage organisation and the process of expansion and absorption of sections of the Dinka by the Nuer must be interpreted with reference to a third factor, the most determinant, which is the development of pastoral production in Nuer society. Unfortunately, we lack precise information on this pastoral production. The discussion which follows thus owes more to a reconstruction of the data than to empirical analysis. Nonetheless, one thing is certain: ecological factors do not determine the social and economic organisation as such but only to the extent to which these factors constitute factors of production.

Southall (1976: 472) has remarked that from an ecological point of view, little distinguishes the eastern Nuer, who have abundant pasturage above the level of flooding and relatively large herds, from the western Dinka who are less rich in cattle. However, the Dinka do not have as large wet season villages as the Nuer nor do they, of course, have the same type of social organisation. If we consider the ecological conditions as they relate to pastoral production, we can establish something of a continuum: 'forest' Dinka/western Dinka/ western Nuer/eastern Nuer, which would represent not only different levels of pastoral production but also different modes of social and economic organisation. Nonetheless, these different forms of social organisation share certain structural characteristics.

Unfortunately, I do not have the space to develop this hypothesis, which is admittedly even more speculative than the preceding one. I will content myself with returning to the questions that have been raised concerning the development of segmentary lineage organisation among the eastern Nuer. One can note that the widening of the patterns of alliance, based both on the rules of complementary opposition between segments and the extension of the patrilineal genealogy, is accompanied by two processes. These processes are, on the one hand, a territorial expansion characterised by the absorption of numerous Dinka elements at the periphery and, on the other, an increasing differentiation of the relative status of lineages (i.e. aristocratic lineages versus other lineages) and of individuals (i.e. 'bulls' — the dominant personalities in the villages). Can we establish a relationship between these different factors and the development of pastoral production?

Evans-Pritchard's analysis (1968) of Nuer society placed emphasis on its segmentary structure, linking the differentiation into aristocratic lineages to the maintenance of equilibrium in territorial relations and insisting on the relative character of this distinction. (That is, a lineage might be aristocratic in one village and not so in another.) Thus, according to Evans-Pritchard, far from being in conflict with the rules of complementary opposition and the structural relativity of segments, the development of aristocratic lineages aided the segmentary lineage system to function effectively by furnishing an armature for local communities. However, although I do not deny the relative nature of the stratification of Nuer lineages and its importance in articulating the descent group structure with the territorial structure in a society where the two structures do not coincide (because local groups are not descent groups), I do not believe that one can reduce our explanation of this phenomenon to that function alone. In my opinion, the differentiation of the status of lineages can also be linked with patterns of differentiation in the status of individuals which itself is more clearly based on socioeconomic factors.

K. Gough (1971) has shown that there is a relation between the position of the 'bulls' and that of the dominant lineages whether they be aristocratic or simply attached to aristocratic lineages. The virtually complete exogamy of local groups, which is due to the extension of matrimonial prohibitions and the tendency for subordinate lineages to be linked with dominant lineages via cognatic or affinal ties, tends to define the field of economic and political competition within the dominant lineages. Glickman (1971) develops this line of analysis by showing that the debts contracted via the circulation of cattle as marriage prestations tend to reinforce this inequality, in particular at the expense of the most recent immigrants who are considered to be strangers to the local community. Thus, the factors which underlie this inequality of status are the unequal capacities of individuals to accumulate cattle as well as the tendencies for social mobility, which determine the different degrees of access to the land. (However, no producer is totally excluded from access, not even the Dinka 'captives' who have become integrated into Nuer society.) Competition for positions of power is based on mobilisation of kinship networks and, to a certain extent, on the accumulation of cattle which itself is linked with and phrased in terms of territorial control. This territorial control, in turn, reinforces the already marked agnatic bias within the society. This bias is expressed by patrilineal descent, by patrilocality, as well as by the fact that the cattle employed in marriage transactions are obtained through the father-son relationship (Glickman 1971: 307). When they are not able to establish themselves as 'bulls', members of aristocratic lineages have the tendency to create new residential groupings. The genealogical structure thus tends to expand at the level of major groupings. As a consequence of the emphasis on agnation which determines both territorial rights and marriage possibilities, aristocratic lineages prefer to use their women in the role of men to reinforce their local positions of power (Gough 1971).

The system is not rigid: certain subordinate lineages, who have the status of 'daughter's children', may supplant aristocratic lineages and gain in power by acquiring the position of 'bulls' in their local groups. In this sense, it is clear that the social dynamics of Nuer society cannot be reduced merely to a segmentary equilibrium but rather express in kinship terms the underlying tendency toward unequal accumulation, especially of cattle. Certain subordinate lineages come to be integrated into dominant lineages via the women of the latter and are able to enter into the competitive arena. Others retain their 'stranger' status.

The suppleness of Nuer kinship rules permits these continual adjustments, with the segmentary lineage structure being the overall summing up of this phenomenon. Genealogical and marital alliance structures expand and call into question the rules of complementary opposition, thus expressing the development of competition within Nuer society. This competition in turn is the result of an expanding pastoral production. Because of the tendency to integrate collateral 'daughter's children's' lineages into the genealogical structure, with concomitant formation of new local groups, this process of internal differentiation is expressed in a more and more extensive segmentary structure. The rules of complementary opposition function to permit cooperation and effective alliances between different local groups.

However, this transformation leads as well to a constant reduction of this differentiation and calls into question the position of locally powerful individuals. In the process of the foundation of new local groupings, the opening up of new pastures and the integration of Dinka 'strangers', the Nuer segmentary lineage system has a tendency to export to its peripheries the contradictions which the increasing economic and social differentiation produces within the system. The expansion which accompanies increasing production, internal differentiation, and the development of segmentary lineage organisation is therefore in no way a territorial conquest; it is rather a redistribution in space of internal political and economic competition. This competition, which initially manifests itself at the level of each autonomous producer, develops within lineages and stimulates the emergence of inequalities of status among them. This status differentiation, on the one hand, is constantly counteracted by the process of territorial expansion which tends to restore, for each producer, the possibility of equal access to the communally held means of production.

This review of the literature on the Nuer and Dinka has no pretensions of being exhaustive, but it does raise a number of more general points: (1) Segmentary lineage organisation can be seen to emerge under certain conditions, among which must be included a number of ecological factors. (2) These situations are accompanied by processes of social and economic differentiation, which do not eventuate in stable hierarchies, however, as a result of the structural characteristics of segmentary lineage organisation. (3) The development of a specific social structure founded on unilineal descent is rendered possible by the special form of the community and territorial organisation which, in turn, owe their form to the ecological conditions. The determining feature of the system, however, is the mode of real appropriation by the producers themselves of the territorially dispersed, collective resourses, that is to say, the organisation of production. (4) Contrary to the postulate which is at the heart of classic social anthropological work on the subject, the distinction between unilineal descent versus the rest of kinship relations, which among the Nuer is expressed as agnatic descent (*buth*) versus cognatic kinship (*mar*), is a functional rather than a structural distinction. This is also the case as regards the distinction between territorial segmentation and lineage segmentation which appears as a consequence of the preceding distinction. I will now expand on these two last points in order to clarify their theoretical implications before returning to the matter of the generality of my preceding conclusions.

Territory and Kinship

Among the Nuer as among all other segmentary lineage societies, the economic content of kinship relations must be examined along with their other functional and structural characteristics. Evans-Pritchard himself (1935) sometimes recognised the need for a more comprehensive point of view.

> Space destroys the sense of unity among kinsmen for kinship is recognised not as a mere genealogical relationship but as a system of reciprocal social service. Your kin are those who help you to pasture your cattle, to marry your daughters, to avenge your dead, and so on. Effective kinship is measured by social obligations and these can only exist when the kin live together (cited by Glickman 1971: 316).

From this perspective, the distinction between *buth* and *mar*, agnation versus cognatic kinship, is linked to a more fundamental distinction in pastoral and semi-pastoral societies between territorial organisation and residential organisation — a distinction which is the key for the next section of my paper.

I have already indicated that in such societies, the appropriation of territory, conceived of as an ensemble of natural resources that have been transformed to a limited extent by human effort, is of a collective nature. Given the environmental characteristics (seasonal or annual climatic variations and spatial dispersion of different types of pasturage and of water points) and technological characteristics (appropriation of these natural resources by means of mobile herds), each producer must have access to all or a part of the collectively appropriated territory. In segmentary lineage societies, these access rights are defined, by and large, by unilineal descent which determines the membership of each producer in the pastoral community. This phenomenon is what I shall term 'the rule of abstract property'.

The more pastoral the production in a society, the more unstable are its residential groupings. These residential structures are inseparable from the different forms of cooperation between herdsmen, given the nature of pastoral work. These structures act to compensate for insufficiencies of herds in certain production units as well as, in the opposite case, for insufficiencies of labour in units possessing large numbers of stock. It is also within the residential group that pastoral work is effectively carried out and that communally held resources are really appropriated. (Real appropriation is here being contrasted to abstract property relations.) These residential structures often bring into play the totality of kinship relations including affinal and cognatic relations and not simply relations of unilineal descent; within Nuer local groups, social relations are organised in terms of cognatic and affinal relations (*mar*) and not solely in terms of agnation (*buth*).

In numerous segmentary lineage societies we find the same opposition between access to territory (unilineal descent) and residence (determined by the totality of kinship relations) but residential structures vary greatly between societies. We can take as an example here three east African societies which are particularly oriented toward camel pastoralism.

The residential organisation of the Somali (Lewis 1961) is structured, in part, along agnatic lines. The younger members of the society, grouped according to their agnatic relations, tend the camel herds separately. These herds must move long distances and can make use of pastures which are located at considerable distances from water sources. The remainder of the family, along with the small stock, live in more permanent encampments which are composed of both agnates and affines.

On the other hand, among the Gabra pastoralists, a Galla group living in the north of Kenya who have been studied by Torry (1976), affinal relations determine almost exclusively the patterns of residential grouping. The obligation of a young husband to live in his wife's encampment for one year after marriage and to work on behalf of his affines leads to constant relocations by whole families and favours the constitution of residential groups composed of brothers and brothers-in-law. These groups form cooperative associations for the purpose of herding and utilisation of watering points. The small number of waterholes during the dry season necessitates a rigorous organisation of cattle watering, which in turn implies a large number of encampments. Here again, the affinal relations which exist between residential units permit this wide-ranging cooperation:

> it is through the affinally structured framework of the local communities that the high degree of collaboration associated with watering and re-location operations is achieved (Torry 1976: 281).

Finally, the Kababish of Sudan content themselves with a family-based division of labour for the purpose of herding their different kinds of stock. They also frequently use hired labour for this purpose. Their residential units are unstable and of a totally composite composition (Asad 1970).

These differences clearly show that although unilineal descent may define abstract property rights over natural resources, this is not necessarily the case with regard to the real appropriation of these resources. Indeed, the cases where residential groupings are actually constituted on the basis of unilineal descent, in certain Bedouin societies for example, appear to be the exception rather than the rule.

The diverse nature of the empirical data poses a certain number of problems. The first, which will be considered more fully in the following section, can be briefly summarised as follows. A major part of anthropological theory concerning segmentary lineage structure is based on the notion that social groups defined by unilineal descent are corporate groups. If it is only in exceptional cases that these corporate groups define the real conditions of appropriation of communally held means of production, since the herds which are the principal means of production are usually appropriated and possessed at the family level, one is led to question the true significance of the corporateness of unilineal descent groups.

The second problem is even more general. In segmentary lineage societies, can it be said that relations of unilineal descent display a certain special structural significance in comparison with other forms of kinship relation? To pose this question is to query the pertinence of the whole model which has been constructed on the basis of such a structural distinction and in this regard, I would adopt the position of Dumont in his analysis of Evans-Pritchard's work,

> le fait de présenter la parenté non agnatique et la parenté mythique agnatique ou non, comme de simples moyens pour établir les relations entre groupes agnatiques et territoriaux ne laisse pas d'apparaître comme arbitraire. C'est dans une grande mesure le résultat d'une accentuation privilégiée des groupes qui relègue au second plan les relations entre groupes quand elles ne sont ni territoriales ni agnatiques (Dumont 1971: 68).

This redefinition of the totality of kinship relations in terms of unilineal descent has, in fact, always been a stumbling block in segmentary lineage theory and has been the origin of such concepts as 'complementary filiation' to explain data that do not fit into the concept of unilineal descent.

Rubel's attempt (1969) to explain the importance of affinal relations in constituting residential groups in pastoral societies seems to me to share the same difficulty. Rubel notes that among the exclusively camel herding Rwala Bedouin, the structure of the encampment is based on that of the agnatic grouping while among groups such as the Bedouin of Cyrenaica and, to an even greater extent, among the Somali who herd both camels and small stock simultaneously, the encampments include a high percentage of affines. The author concludes that mobilisation of affinal relations in segmentary lineage societies is linked with pastoralism based on mixed stock herding which, in turn, demands access to a diversified range of environments.

In fact, the data demand a more refined treatment. (For example, the Rgeybat of the western Sahara who are exclusively camel herders do not base their residential groupings exclusively on agnatic relationships.) It is true that, in general, affinal relations are used by these societies to assist them in utilising their pasture resources in a more differentiated way. But Rubel is not content with this observation. Since access rights to community resources are defined by unilineal descent at the level of abstract property, the real appropriation of these resources can also only be defined by descent. As a result, affinal relations permit an individual to benefit from the rights of other unilineal descent groups. Recourse to ecological and technological explanations of this fact is simply designed to preserve the theoretical model and one of its postulates — the separate structural character of descent from other kinship relations.

Before discussing further the conditions under which descent groups function as corporate groups and the related issue of the development of segmentary lineage organisation, I feel I should go beyond these critical remarks and furnish some indication of the way I see the relation between territorial organisation, residential organisation, and kinship relations. Since I have developed this analysis at length elsewhere (Bonte 1974; 1977; 1978a), I will simply summarise my principal conclusions.

The distinction between territory seen as the location of the real appropriation of the totality of natural resources used by the community and territory as it is formally appropriated by a human group holding abstract property rights seems to me to be a fundamental one. At one extreme, territory understood in the second sense may not exist in some pastoral societies. One of the best known examples of this situation is that of the Wodaabe Fulani studied by Dupire (1970). These pastoral Fulani communities, which are organised into unstable migratory groups (in order to gain access to new pastures and to escape subordination to other sedentary and nomadic societies), are settled on pasture lands which do not belong to them. Even the usage of wells which they have dug themselves may be precarious and is often subject to dispute as a result. During their annual transhumance, communal appropriation of natural resources becomes confused with the real appropriation of natural resources due to the shifting combination of domestic production units in the course of their migrations and political competitions. Fulani social organisation takes this special situation into account,

> La dynamique résidentielle est à la base de la formation et de l'évolution des unités sociales et politiques: des familles qui vivent ensemble finissent par fusionner en une lignée (*wuro*); une branche de lignage, une famille ou un individu qui ont choisi de résider en dehors de leur fraction politique, modifient à la longue leur allégeance et, si la distance est plus grande (résidence avec un autre groupe migratoire), leur appartenance lignagère en même temps que politique (Dupire 1970: 240–41).

Other examples of pastoral communities which do not hold territorial rights include many of the Mediterranean pastoralists, such as the Yorük of southern Turkey studied by Bates (1973). However, the Fulani example is of particular interest. Even though a communal territory does not exist, there is still the need for a collective appropriation of natural resources. In the Fulani case, as in many others, unilineal descent fixes the social limits of the community. But descent relations are frequently adjusted according to the patterns of real appropriation of the collectively used resources, and the patterns of cooperation and residential organisation around the wells. It is natural that this system does not develop into one of segmentary lineages since the allegiances of individuals to lineages and of lineages to larger groupings are constantly changing. In addition, unilineal descent does not exclusively define the limits of communities. These limits are also established by the ceremonial activities which take place at several levels of community organisation (*worso* at the level of the migratory group, *gere(w)ol* at the level of maximal lineages).

It must be explained that territory, seen as a set of collectively exploited resources, still has a real existence even in the absence of a corporate territorial organisation to define the community. In order to understand this special characteristic of territorial organisation, it is necessary to avoid two pitfalls. The first is illustrated by the work of E. Marx (1977) concerning the notion of the tribe in pastoral societies. This author considers that the tribe as it exists in pastoral societies in the Middle East is, above all, a 'unit of subsistence' which exploits a particular zone of pasturage and particular water points. It is 'mainly a territorial ecological organization' according to Marx (1977: 348). But Marx reduces the other aspects of tribal organisation to this same point; for example, the genealogical structure is understood as a direct translation of the rules of access to land.

The opposite error is illustrated by certain marxist studies concerning the 'domestic mode of production' or the 'lineage mode of production'. Because the real appropriation of the communal means of production (the territory) takes place exclusively at the level of domestic groups or lineage groups, these authors conclude that appropriation at the community level does not exist. According to Rey (1975), the more strongly tribal property is affirmed to exist, the less there is real appropriation at the level of the tribal estate. This paradox is linked to the fact that the existence of tribal property is not produced by the need to have access to a territorial domain, since land is in abundance, but by the reciprocity which is established between lineages and, more particularly, by the relations between the lineage elders. It is these elders who control the social circulation of women and men – that is to say, in the last analysis, the labour force. Such an analysis does not take into account the necessity, which is built into the conditions of production and reproduction of each domestic group or lineage, of having access to all or part of the territory which is appropriated at the community level. This necessity is evident among nomadic pastoralists, but it is also present in mobile agricultural societies due to the fact of fallowing and the shifting cultivation of fields. From this point of view, one may ask oneself if the notion of abundant land in these societies is not a very relative one. Even if it is very, or even excessively, abundant in a normal season, available land may be insufficient, for example, during drought periods. It is indispensable in calculating the degree of abundance of land to take all aspects of the ecosystem into account. In fact, as the example of the Wodaabe Fulani shows, it is less the case of property without real appropriation that must be explained (as Rey thinks) than the case of collective appropriation without territorial property. However, the second aspect of Rey's analysis is correct: property corresponds here to the social form of community relations more than to a corporate-territorial organisation. In order to interpret this type of social organisation, I have introduced a conceptual distinction between the community engaged in production and the community 'in itself' (Bonte 1978a). This distinction permits us to take into account the fact that, whereas in the course of the production process each producer is effectively appropriating the communal means of production and in his role as an autonomous producer is thus proprietor (on behalf of the community) of these means of production (the community appearing as an association of these independent producers), this effective appropriation will be hidden and will appear to be the result of formal appropriation (a matter of abstract property) at the community level.

The characteristics of this association may be determined according to the rules, explicit or implicit, which govern the utilisation (real appropriation) of these collective resources. Although each producer is autonomous for all practical purposes, access to pasture and water cannot be accomplished in an anarchic manner. To the contrary, this access must be arranged in such a way that non-exclusive utilisation of these resources is obtained. A good example of this is furnished by the organisation of usage of summer pastures among the Ayt Atta of Morocco (Lefebure 1978). At the highest level of segmentation, the Ayt Atta are organised into five sections which are themselves divided into tribes (*taqbilt*). Defence and utilisation of pasturage is organised at this level. Every year, a head of each grouping is elected who has charge of all the Ayt Atta pastures. The exploitation of the summer pastures in the mountains is particularly carefully organised; certain of the pastures (*igdoulan*) are put off limits and surveyed by a guardian. Rights of access to pasturage between competing groups and individuals are thus precisely controlled to the point that camp locations are allocated by the community. Among the Baxtyari of Iran, this organisation of pasture utilisation and the usage of trans-humance routes is regulated with even more rigour – if need be by means of written agreements which are recognised by the whole group (Digard 1978). Among the Gabra mentioned earlier, it is especially the usage of water resources which must be strictly organised. Torry mentions that decisions concerning the watering of stock are taken at the level of neighbourhood assemblies which may encompass representatives from as many as thirty or more encampments. These decisions are backed up by force if necessary (Torry 1976).

In pastoral production, therefore, it is on the basis of more or less voluntary and equal forms of association that the effective utilisation of collective resources is determined. The rules which govern this association, just as those governing the type case of residential organisation, are phrased in terms of kinship without descent relations being particularly singled out. One may consider that the kinship relations which determine the form of cooperation in work, which control the mode of appropriation of the communal and domestic means of production, and which determine the way in which production is shared, function as relations of production. But kinship relations and, more particularly, unilineal descent simultaneously fulfil another function: they determine the conditions of formal appropriation of the territory (abstract property) which appears fixed for each producer on the basis of his genealogical position. This is what I have termed the 'fetishism of kinship' (Bonte 1977) in discussing the way in which the functions of kinship as relations of production are hidden by this particular function of descent. The manner of determining territorial organisation via kinship relations has one immediate consequence. Since the appropriation of communal resources (and the appropriation of stock) takes place in practice at the level of domestic groups, either isolated or in cooperation with others, this continually calls into question the equality of access of these various groups whose production tends to be unequal. This equivalence must be structurally established, via the genealogical structure, at a level which is not that of the real appropriation of the means of production but which appears as their precondition.

One can now better understand how the theoretical model of the segmentary lineage could be developed in such cases. It rests on a double error: (1) it transforms the pluri-functional definition of kinship, which is particularly determined by the organisation of production, into a structural distinction within kinship itself, and (2) it accepts the apparent determination of the social organisation by unilineal descent. Thus, this theory attributes the characteristics of corporate groups to social groups defined by descent despite the constant counterevidence of the facts which attest to the exceptional character of such a situation.

Analysis of Corporate Groups

A corporate group holds rights over material or immaterial possessions (an estate), in particular over the means of production and work of its members. In the theoretical debates on this subject in anthropology, which have been well summarised by Dumont (1971: 75 *et seq.*), the central issue has been the land tenure rights attributed to the descent group seen as a corporate group. To the extent that this question raises the issue of the link between territorial organisation and kinship relations, it is truly the fundamental one.

In the previous section it has been shown that the automatic attribution of corporate group characteristics, in particular land tenure rights, to a descent group results from a confusion over the function of unilineal descent. Although descent appears to determine the conditions of appropriation of the collectively held means of production, in fact it only defines the rules of abstract property at the community level. I shall attempt in this section to specify several of the conditions under which we can attribute a corporate group character to descent groups.

To begin with, there must exist a set of technological and ecological conditions which permit an allocation of fixed rights of formal appropriation to the different levels of segmentation in the lineage system. Comparison of several Bedouin societies will help to clarify this point.

The Rgeybat pastoralists of the western Sahara live in a region where the interaction of several climates (Mediterranean, Atlantic, and the tropical monsoon) and the diversity of soil types (*reg, erg, hamada*, etc.) render migratory movements totally unpredictable. The light rainfall accentuates the dispersal of pasturage, and each group must have access to the whole of the *trab er-rgeybat* which stretches over nearly 500,000 km². For example, between 1934 and 1944, the encampments of the Sidi Allal section were never located on the same site twice. Because of this fact, territorial segmentation is weakly developed (it is limited to a division of the tribe into two localised sections, one easterly and one westerly) and the rights of tenure are held exlusively by the tribe as a whole. Other groups also pass through the same territory.

The Arabian Bedouin, the Al Murrah, who raise camels on the sandy pastures,of the Rub al Khali (the Empty Quarter), utilised winter pastures all the way to southern Iraq. Their vast territory of 650,000 km² is hardly subdivided despite the existence of some stable groupings around summer wells. A certain stability of grouping is also necessary in order to be able to move from summer to winter pastures because of the military pressure of other tribes. This transhumance is carried out within the framework of lineage organisation (Cole 1975).

The Rwala Bedouin of the Syrian desert are also camel herders. In the summer, they are grouped around oases and near to Damascus and in winter they utilise distant pastures to the south. Their transhumances are very long but follow a relatively stable pattern. Rights to pasture are established at the level of the lineages (*hamula*), the ranks of which are swelled by numerous clients and tributaries. Chelhood (1971: 347) has written in this regard,

> Le collectivisme du désert se situe au niveau de la *hamula* dont les membres peuvent faire paître leur troupeau sur n'importe quel point du lôt commun. Les autres groupes de la tribu n'y sont admis que sur autorisation du *cheykh* en raison de leur parenté, leur voisinage ou leurs bonnes relations avec les possédants en titre.

In the case of the Cyrenaican Bedouin (Peters 1967) and those of the Negev (Marx 1967), rights to pasture are stronger at the level of the lineage. These Bedouin utilise lands close to the sea which are relatively well watered, with regular rains, and rights of access to territory can be attributed to a tribal section or even an individual lineage — the tertiary level in the segmentary system of the Cyrenaican Bedouin.

> Tertiary sections are discrete also in the sense that all the major necessities of daily living are available within their several territories. Each controls its own natural resources, and these resources are the same for all the camel-herding groups — in any given year one may have more of a particular commodity than another, but the difference is in degree and is ephemeral, so that the opportunity for the development of trade relations is absent (Peters 1967: 262–63).

Nonetheless, the need to have access to other parts of the tribal territory which are exploited by different sections of the tribe arises from time to time, in drought years for example. Access is facilitated by the relations maintained between sections, in particular those of affinity. Among the Bedouin of the Negev, the stability of the implantation of the lineage (*khams*) in a portion of the territory is even more accentuated, but it is necessary to keep in mind the policies of fixed settlement and territorial restriction of the Israeli military administration, practices which are reminiscent of those of the Ottoman Empire.

In this way, comparison of different Bedouin groups shows that a certain set of technological and ecological conditions must be present for descent groups to have more or less exclusive rights of appropriation over a part of the communally held territory (that is to say, over the natural resources). But although these conditions are necessary, they are not sufficient. In the Sahelian zone and in the Sudanic regions of Africa, the relative regularity of patterns of transhumance toward the wet season pastures located in the north permits a definition of rights at the level of the lineage. However, neither the Moors of southern Mauretania nor the Baggara of the Sudan (Cunnison 1966) display such an organisation. In both societies, only agricultural lands are the object of lineage or even familial appropriation. It is necessary, indeed, to invoke other conditions of a structural order to explain the development of a lineage as a corporate group. Certain of these structural conditions are external to kinship — for example the form of political organisation which surmounts this segmentary lineage structure since integration into a state will modify its functioning. (I will return to this point in the next section.) Other conditions are related to kinship structures themselves — in particular, to the form of marriage.

Up to now, anthropological theory concerning descent groups has
not distinguished between descent groups which are exogamous and
those in which a marriage partner may be chosen either from inside
or outside the group. On the basis of his Nuer model. Evans-Pritchard
treated these two types of lineage as one when he came to work on
the Bedouin of Cyrenaica (1949). The primacy accorded to unilineal
descent tended to relegate marriage rules to the background. In fact,
of course, these two types of descent groups do not display the same
characteristics, especially with regard to their potential for corporateness. Likewise, they do not have the same capacities for adjustment
of the different factors of production (work/stock/pasturage). The
distribution of these factors of production within a pastoral society
is continually being perturbed by social dynamics, especially those
demographic and ecological variations which are difficult to foresee.

Where rules of lineage exogamy do not exist, descent is formally
defined as unilineal but is in fact undifferentiated. This is the case,
for example, in Bedouin pastoral communities and among the herdsmen of the Middle East and the Maghreb, etc. where marriage is even
contracted by preference within the lineage (with the father's brother's
daughter). Genealogical manipulation and the forgetting of certain
non-agnatic kin permits the members of these societies to adjust
the groupings of heterogenous co-resident kin according to an agnatic
plan. Dupire (1970: especially Ch. 7) has shown how agnatic lineages
are constituted among the Fulani on the basis of co-residence and
intermarriage. Barth, for his part, has analysed how Basseri encampments, which are residential groups of relatively large size in this
society and are constituted on the basis of heterogenous kin
relationships, are constantly evolving toward a structure which
emphasises agnatic descent. This change in the composition of
encampments defines the potential lines of cleavage within the
oulad, the patrilineages holding rights of access to territory (Barth
1964).

Capacities for genealogical manipulation also exist in those
societies where lineages are exogamous — societies whose prototype
is the Nuer case. This genealogical manipulation takes different forms:
manipulation in order to secure the integration of 'stranger' groups
at a high level of segmentation, manipulation of descent relations
such that a uterine relation is redefined as agnatic descent, etc.
However, this manipulation is less efficacious. It appears as if the
supplemental function of descent in these societies, that of
organising marital exchanges between groups, renders more difficult
the ideological function of the lineage structure, which is to hide the

role of kinship relations in production.

In effect, the circumstances under which descent groups constitute corporate groups — circumstances which are linked with a particular set of technological, ecological and structural (kinship) conditions, correspond to a widening of these ideological functions of descent (dissimulation). The function of descent to hide the role of kinship relations in production takes place not at the level of macro-genealogical structure but at the level of each lineage or at least, at some such lower level of segmentation. This fact has specific effects on the functioning of segmentary lineage organisation.

From the point of view of the classic anthropological model developed to interpret segmentary lineage societies, one would expect that the structure would function better the more corporate the descent groups. However, in an important article, Peters (1967) has shown that among the Bedouin of Cyrenaica whose descent groups have a clearly defined territorial basis, the feud is not organised according to the predicted rules of complementary opposition but according to a completely different logic. Full scale competition exists between tertiary segments which are spatially and genealogically close to each other, but relations of political alliance are established between secondary sections different from those specified in the genealogy. This is done with the particular aim of having access to each other's territory in case of need. In no case are alliances established according to the segmentary lineage genealogical schema:

> Regularity in relationships does exist, it is urgently necessary, and it persists; but this regularity is not consistent with a lineage model. Ultimately, feud is a violent form of hostility between corporations which has its source in the competition for proprietory rights in land and water. This competition makes it necessary for groups to combine to prevent the encroachments of others in similar combinations and also to expand their resources whenever the opportunity arises. The significant groups in a discussion of the feud are these power groups, and it is their composition, the shifting alliances within them, the growth and diminution in the power of tertiary sections constituting the combinations which makes the facts of feud intelligible (Peters 1967: 279).

Does this constitute a sufficient reason to reject analyses based on segmentary lineage concepts as a case of an 'indigenous' model being wrongly employed as a sociological model? Salzman (1978) reaches a more subtle conclusion: the difficulties in applying the model to the Bedouin of Cyrenaica are due to the fact that this is not a 'pure' case but one which has been complicated by relations of locality. From this point of view, there would be an incompatibility between the territorial stabilisation of a society and the maintenance of segmentary lineage organisation. This is an interesting observation in relation to my earlier remark concerning the direct relation between this type of organisation and the occurrence of territorial expansion. But, is it necessary to consider, as Salzman does, that the segmentary lineage model based on complementary opposition between segments is 'a social structure in reserve' (Salzman 1968: 69)? It seems to me that it is the very notion of the model which must disappear. It is less a matter of employing a single model (freed of all considerations of 'impurities' and variation) in the analysis of various societies displaying common traits than of defining the different circumstances in which principles of social organisation held in common (such as the role of kinship in production, the 'fetishism of kinship', and the distinction between the community engaged in production and the community 'in itself') develop into different social structures. Certain provisory conclusions in this regard have already been put forward in the course of my analysis of the Nuer case. I will try in this last section to consider the possibility of generalising them in order to understand the development and reproduction of segmentary lineage systems.

The Development and Reproduction of Segmentary Lineage Systems

Mauretania and its neighbouring regions constitute, as in the case of the Nuer and Dinka, a sort of 'laboratory' for the analysis of segmentary lineage societies. The diversity of the observable forms which at many points have much in common, permits one to draw out certain of the factors which determine the development of such societies. I will restrict myself to the discussion of one case, that of the Rgeybat, who lived in parts of the western Sahara where the rains were sufficiently reliable for the cultivation of barley, and formerly practised a mixed economy of agriculture and small stock pastoralism. They shifted to camel pastoralism only in the 19th century and began their conquest of the desert pasturages of the interior which were very suitable to camel husbandry. At the same time, their social structure was transformed. Formerly a subordinate tribe made up of heterogenous elements grouped around a saint named Sidi Ahmed Rgeybi (or rather around his cult), they increased this diversity in the course of their expansion. Numerous stranger groups were integrated as clients while, at the same time, the tribe developed a genealogical structure which placed the tribal sections and the dominant lineages into relations of complementary opposition.

The transformation of the Rgeybat social system took place,

therefore, in response to changes in technology (a dominant and sometimes exclusive reliance on camel husbandry), in ecology (the opening up of vast pasturelands, the *trab-er-rgeybat*, the exploitation of which in the conditions of unpredictable rainfall necessitated great mobility) and in politics (development of clientage relations, absorption of new groups and individuals from the outside, and also the development of slavery). No one of these factors, however, is the determining one just as all of them taken together cannot explain the development of their segmentary lineage organisation.

As in the case of the Nuer, the common denominator of these various change processes in 19th century Rgeybat society is the development of pastoral production, which has been accentuated in this case by the 19th and 20th century wars which allowed the Rgeybat to accumulate a very large camel herd.

Let me now make precise what I mean by the development of pastoral production. In the present circumstance, it is a matter of the development of the productivity of pastoral work, the amount of which can be approximately estimated, all things being equal, by the increase in per capita holdings of stock. The level and variations in the productivity of pastoral work cannot be reduced to technological changes or to ecological fluctuations but are a function of a specific set of social relations (the relations of production) and are accompanied by changes in these social relations. In the Rgeybat case, this variation in productivity has been accompanied by change in kinship relations since these govern the social relations of production. It is from this perspective that one must interpret the accentuation of social differentiation and the simultaneous development of segmentary lineage organisation in this society.

Among the Rgeybat, social differentiation is very clear cut. It takes three forms: (1) The number of slaves is on the increase in the 19th century and in the beginning of the 20th century; this slavery is of the domestic type and reinforces the capacities of the domestic work groups which can thus maintain larger herds. (2) Loans of camels (*meniha*) permit rich herd owners to disperse their herds and avoid the necessity of augmenting the work force of their own domestic groups. Since the reproductive increase of the loaned animals is returned to the lenders, this is also a means of exploiting the labour of the less rich herdsmen. The capacities for accumulation are thus augmented and certain herd owners may own several thousand camels. (3) Finally the clientage relations which are established at the individual or total group level constitute another form of social differentiation. Such relations can be established via a sacrifice (*targiba*) carried out in front of the tent of one's protector (by hamstringing the animal). These relations imply a series of voluntary and irregular prestations and especially, participation in *diya* payments (blood money). Such relations are phrased ideologically in terms of protection. In fact, the very large groups which may become clients are completely able to assure their own protection, and the clientage relation is rather a mode of integration into the segmentary system.

The clientage relation among the Rgeybat also illustrates clearly the contradictory framework within which the segmentary lineage organisation develops. The increasing inequality between producers is the consequence of the increase in productivity of work (unequal accumulation) which leads to the establishment of forms of social relations such as the unequal loan or the clientage relation. The integration of clients into the segmentary structure, however, assures them of equal access to the collectively held resources and prevents the transformation of this unequal accumulation into a more definitive inequality of access to the means of production (except in the case of slavery). In this respect, there are comparable features between the aristocratic relation among the Nuer and the clientage relation among the Rgeybat. One can multiply these examples. Among the Rwala Bedouin, the lineage (*hamula*) surrounded by its clients and tributaries and led by the *cheykh*, has a proprietary right over a tract of common territory (Chelhod 1971). In other Bedouin groups of north Arabia, the residential unit is a group of patrilineages with one lineage among them dominant. The chieftaincy (*dira*) thus constituted has traditional rights over a pasturage zone. The *rubae*, which is the lowest level of territorial access among the Bedouin of the Negev, is generally composed of a Bedouin lineage (*khams*) through which other exterior 'peasant' groups (*fallahin*) and slaves gain access to land (Marx 1967). (These *fallahin* have, in fact, the same life style as the Bedouin but are not genealogically attached to them.) In the same manner, the Cyrenaican Bedouin studied by Peters are composed of noble groups to which are attached numerous clients. Thus, at one extreme, segmentary lineage organisation has the effect of justifying and reproducing the domination of one part of a society by another.

In fact, although one can understand how segmentary lineage organisation can be linked to the development of pastoral production and the social differentiation which accompanies it, one cannot always understand the 'why' of this process. To do so, we must now think back to the analysis of the Nuer which we began earlier since it does have a certain generality. I noted in the first section of the paper that the development of segmentary lineage organisation in Nuer society tended to export to its peripheries the contradictions produced by increasing social and economic differentiation. The most important of these contradictions was the same one that we have also just identified among the Rgeybat, namely the more and more unequal pattern of accumulation and the necessarily equal access of each producer to the collectively held resources. This contradiction is made manifest in the development of clientage relations through which 'foreign' elements are incorporated into the tribe. This incorporation, which becomes more frequent during a period of territorial expansion, is carried out in an unequal fashion symbolised by the *targiba* sacrifice. But especially in cases where they are numerous and powerful, such clients tend progressively to integrate themselves into the structure of the segmentary system in order to gain direct access (not through clientage) to the collectively utilised resources. This process of incorporation is all the more easily and rapidly accomplished because the tribe is numerous and widely dispersed over a vast territory, because the tribal segments have practically no relations between themselves and because the sections preoccupy themselves only with that portion of the tribal genealogy which concerns them directly. The tensions which result from this contradiction and the practices which permit them to be reduced are also expressed in warfare, which has an important role in this society of warriors. War multiplies the opportunities for unequal accumulation, in particular due to practices such as silent partnership (*^catila*). Thanks to this custom, a rich herdsman who furnishes military equipment to a poor warrior obtains a major part of the booty, but this is also an opportunity for poor young warriors, who may have no possibilities of inheriting stock, to gain access to animals. But warfare also produces both a social as well as a geographical redistribution of the camel herds and thus militates against overstocking of the best pastures. In one sense, warfare reinforces the capacity for unequal accumulation in society but in the last analysis, it also effects a redistribution at the level of community or inter-community relations.

The extension of the genealogical structure and the introduction of rules of complementary opposition thus permit the social tensions which result from the growing and unequal production to be exported to the periphery of the system through the opening of new pastures and the redistribution of opportunities for acquisition of stock. These processes also permit the maintenance of the patterns of alliance and cooperation necessary for the development of pastoral production itself. For this reason, segmentary lineage organisation prevents these internal contradictions from producing more unequal forms of society. Under different circumstances, as for example among the southern neighbours of the Rgeybat, the Moors of Mauretania, and among other segmentary Saharan tribes, class relations can emerge. In the last analysis, Sahlins' study (1961) does properly identify expansion as the circumstance under which segmentary lineage organisation has a tendency to develop, but he makes the error of identifying this expansion as a feature of this organisation itself. This error results from his unacceptable evolutionist approach. I have here tried to show that, in fact, the development of this mode of social organisation is the result of a process of economic growth in these societies where the community structure is based on 'kinship fetishism', the role of kinship in production, etc. Far from being its cause, segmentary lineage organisation is rather the form which the expansion takes: it transforms economic growth into a process of peripheral expansion which then blocks the development of the con-traditions internal to the society. This change is rendered possible by the functioning of unilineal descent. Under its guise, expansion appears as a process of collateral ramification and an extension of the genealogical structure. But in more general terms, the emergence of segmentary lineage organisation expresses the fact that kinship relations are functioning as relations of production. The social changes produced by an economic cause consequently take the form of structural changes within the kinship system itself.

One must return now, at the end of this study of the conditions of development of segmentary lineage organisation, to the matter of the role of ecological factors in this process. These factors intervene, in my opinion, at the level of the conditions of economic growth itself.

One may note that expansionary movements of segmentary societies tend to emerge from certain geographical zones which seem to be the consistent sources of such movements. Thus in the western Sahara, the regions of Tiris-Zemmur constitute a centre for the successive waves of tribal occupation of Mauretania. In the 10th century, tribal groups were settled there which were to give rise to the Almoravids. In the 15th and 16th centuries, it was the turn of the Arab *hassan* tribes who were to establish, in an identical fashion, their political dominion further south and contribute to the con-stitution of the Moorish emirates. Finally, in the 19th century, we have the Rgeybat expansion which I have already discussed. It is noteworthy that these populations, which possessed a segmentary lineage organisation when they were settled at Tiris and Zemmur, experienced a rapid development of social and political stratification at the end of their expansionary process. This fact confirms our hypothesis that this expansion permits such societies to resolve their internal contradictions which, when expansion is no longer possible, then produce a different form of social organisation. When speaking of these zones in which expansionist movements have arisen, we must not forget the deserts of the interior of Arabia which may have been the point of departure of the movements of the Semitic pastoralists which were carried out over millenia (Moscati 1959). In east Africa, it may well be the case that the region of Lake Rudolph was a source zone of the same order not only for segmentary tribes but also for those organised on the basis of age grades. There is also a centre of development and spread of the Nilotic peoples in the present zone of settlement of the Nuer and Dinka (Ogot 1964: 288).

We can note immediately in our enumeration of these regions that they are, in a paradoxical manner, among the most unfavourable from an environmental point of view. This is the case with the annually flooded country of the Nuer as well as in the desert zones with their light and unpredictable rainfall. In fact, this situation is paradoxical only to the extent that there has been a confusion in most anthropological works between two forms of productivity. On the one hand, there is the productivity of the ecosystem, which is best distinguished by using the term 'yield' and which can be quantified by adding together all the material exchanges which take place in the environment. On the other hand, there is the productivity of work, which is calculated by relating the quantity of social work to the production output. These two types of productivity are, of course, not without relation to each other. The productivity of the ecosystem is one of the factors which determines the productivity of work. However, they are not the same thing since the productivity of work is also determined by the form of social relations in the context of which the work is carried out. Confusion of the two concepts, which has the effect of reducing the economy to the ecology (which is often the tendency in present-day anthropological works), is the result of the false idea that the system of production always functions to achieve an optimal utilisation of resources. In fact, production cannot be reduced to this notion of adaptation and of equilbrated utilisation of the environment. In most societies, including capitalist societies, men do not use all available productive capacity. The existence of variations of productivity expresses the specific social conditions of production in any one society (for example the crises and the continual increases in the productive forces in capitalist societies). Moreover, variations in the productivity of the ecosystem and variations in the productivity of work are not necessarily exactly parallel. Thus, the productivity of the ecosystem may decrease at the same time as the productivity of work is increasing, as in the case of overexploitation of the environment. On the other hand, the productivity of work may decrease while that of the ecosystem remains stable as a result of the marginal yield or of increasing population.

From this point of view, it is possible to understand how un-favourable environments (which have a relatively low productivity) permit or even impose a greater specialisation in pastoralism which in turn favours increasing productivity in pastoral work. These environments also undergo brutal variations in production which lead to a rapid increase in the productivity of pastoral work. This increase is proportional to the reduction of the initial level of the produc-tivity of this work (through phenomena such as warfare, droughts, epidemics, etc.). It seems to me that these are the conditions which permit a rapid increase in production in societies that are situated in relatively unfavourable environments and which constitute the explanation of the process of expansion on which this article has focused.

Conclusions

The conclusions of this article must remain very open and provisional given that its particular purpose has been to present some new working hypotheses and to reorient research in an area in which previous theories had tended to become overly rigid. The principal points of my argument, however, can be summed up as follows:

1) The emergence of segmentary lineage organisation, in the pastoral and semi-pastoral societies which I have discussed here, is linked to the development of several structural principles. This form of organisation develops in societies where kinship relations control the social relations of production. This organisation is linked with an extension of the role of unilineal descent in these societies, which acts to hide the role of kinship in production and to give the appearance that the relations of production are determined by the structural form of kinship. The functioning of kinship relations expresses a particular form of community organisation which operates in such a way that territorial organisation, as a collective appropriation of natural resources, is implicated in the production process (wherein each producer appropriates collectively held resources for himself). In addition, in its social form (in this case, as the abstract property of a human group defined by descent), territorial organisation appears

as a prerequisite of all production.

2) Aside from the development of these structural principles, the development of segmentary lineage organisation is linked with specific circumstances of economic growth which lead to greater and greater social and economic differentiation. The very functioning of the community organisation has the effect that this economic growth (an increase in the productivity of social work) is manifested in the form of an unequal accumulation among producers. The development of segmentary lineage structures prevents this differentiation from transforming itself into perennially unequal relations and exports to the peripheries of the system the contradictions which result from it. This mode of organisation thus constitutes the very form which the expansion process takes. It is therefore a transitory mode of social organisation which develops and reproduces itself in situations of economic growth. It is also able to reproduce itself in other situations, however, as a form of territorial administration or as the structure of segmentary pastoral societies and eventually develop itself in the context of class relations (Bonte 1978b).

3) Ecological factors contribute to the creation of circumstances which are favourable to expansion. Paradoxically, the most unfavourable environments appear to be source areas for expansion. In fact, they are conducive to rapid developments in the productivity of pastoral work and to the phenomenon of accelerated growth which itself is favourable for the development of segmentary lineage organisation. By confusing the productivity of work with that of the ecosystem (the yield), the majority of studies have been unable to resolve this apparent paradox correctly.

4) Certain ecological conditions, such as a diversification of natural resources liable to appropriation by relatively small human groups, as well as other conditions which are linked to the very nature of kinship and marriage structures, lead to the definition of a descent group as a corporate group. This only really occurs in the type case, in which the role of kinship in ideologically obscuring the underlying importance of the relations of production is especially effective. To the extent that the patterns of cooperation and alliance which are brought into play in the context of the real appropriation of collectively held resources are in contradiction with these rules, they make the functioning of the rules of complementary opposition defining the segmentary lineage organisation even more difficult.

5) The debate concerning the corporate nature of unilineal descent groups points to one of the stumbling blocks for anthropological theory, which tends to generalise from a type case at the risk of con-

tradiction by the observable facts (if it does not simply content itself with describing social diversity without trying to explain it). This tendency is related to a double error which traditional anthropology has made from the very outset. This theory has transformed what is in fact a functional distinction into a structural one in opposing descent to other forms of kinship relation. It has also developed a concept of equilibrium which tends to exclude from consideration a whole series of factors which are linked to the development of an overarching genealogical structure and to the rules of complementary opposition, in particular the process of social differentiation. From this point on, any empirically observable variations in the structure are conceived of as the result of a dysfunction in the 'pure' model of segmentary society.

6) Because of this situation, references to ecology in order to justify variations in the model are not fortuitous, and the confusion between the productivity of work and that of the ecosystem is not simply a factual error. The notion of adaptation, borrowed from ecology, is used to explain the variations in the 'pure' model. It also justifies the equilibrium imputed to the sociological model under stable conditions of utilisation of the environment. Finally, by reducing the functioning and transformation of the societies under study to a matter of the adaptive character of the social system, it permits a limitation of the role of economic determinants. In fact, the process of adaptation and the equilibrated relationship between a social system and its environment are not the final result of production. Ecological factors must be reinterpreted as factors of production.

7) The present discussion has focused on a particular technological and ecological framework, that of pastoral and semi-pastoral societies. These societies furnish us with a typical example of a certain form of community organisation which is, however, not restricted to societies with such a material base. The processes which I have discussed here are also at work in societies possessing the same social structure yet which practise extensive agriculture; this is the case, for example, among the Tiv who are a classic instance of a segmentary lineage society. We can also see that, here again, we are dealing with a society which experienced a rapid expansion which was carried out not in the form of a territorial conquest but rather within the framework of a constantly expanding segmentary organisation.

References

Asad, T. 1970. *The Kababish Arab: power, authority and consent in a nomadic tribe.* London: C. Hurst.

Barth, F. 1964. *Nomads of South Persia.* Oslo: Universitets forlaget.

Bates, D. 1973. *Nomads and farmers: a study of the Yorük of Southeastern Turkey* (Univ. of Michigan Anthrop. Papers No. 52). Ann Arbor: University of Michigan Press.

Bonte, P. 1973. La formule technique du pastoralisme nomad: Etudes sur les sociétés de pasteurs nomades, 1. Sur l'organisation technique et économique. *Cahiers du CERM* **109**, 6–32.

——— 1974. Etudes sur les sociétés de pasteurs nomades, 2. Organisation economique et sociale des pasteurs d'Afrique Orientale. *Cahiers du CERM* **110**.

——— 1977. Classes et parenté dans les sociétés segmentaires. *Dialectiques* 21, 103–115.

——— 1978a. Non-stratified social formations among pastoral nomads. In *The evolution of social systems*, J. Friedman and M. Rowlands (eds.). Duckworth: London.

——— 1978b. Segmentarité et pouvoir chez les éleveurs nomades saheriens: éléments d'une problématique. In *Le Pastoralisme Nomade: Production Pastorale et Société.* Cambridge: University Press.

Chelhod, J. 1971. *Le droit dans la société bedouine.* Paris: M. Riviere.

Cole, D.P. 1975. *Nomads of the nomads: the Al Murrah bedouin of the Empty Quarter.* Chicago: Aldine.

Cunninson, I. 1966. *Baggara Arabs.* Oxford: Clarendon Press.

Digard, J.P. 1978. De la necessité et des inconvenients pour un Baxtyari d'être baxtyari. Utilisation de l'espace et pouvoir politique chez des pasteurs nomades d'Iran. In *Le pastoralisme nomade: production pastorale et société.* Cambridge: University Press.

Dumont, L. 1971. *Introduction à deux théories d'anthropologie sociale.* Paris: Mouton.

Dupire, 1970. *Organisation sociale des Peuls.* Paris: Plon.

Evans-Pritchard, E.E. 1935. The Nuer Tribe and Clan. *Sudan Notes and Records* **18**.

——— 1949. *The Sanusi of Cyrenaica.* Oxford: University Press.

——— 1968. *Les Nuer.* Paris: Gallimard (English edition 1940, Oxford University Press).

Glickman, M. 1971. Kinship and credit among the Nuer. *Africa* **41**, 306–19.

——— 1972. The Nuer and the Dinka: a further note. *Man (U.S.)* 7, 586–94.

Gough, K. 1971. Nuer kinship: a reexamination. In *The translation of culture: essays to E.E. Evans-Pritchard*, T.O. Beidelman (ed.). London: Tavistock.

Leach, E. 1961. *Rethinking anthropology.* London: Athlone Press.

Lefebure, C. 1978. Accès aux ressources collectives et structures sociales: l'estivage chez les Ayt Atta. In *Le pastoralisme nomade: production pastorale et société.* Cambridge: University Press.

Lewis, I. 1961. *A pastoral democracy.* Oxford: University Press.

Marx, E. 1967. *Bedouin of the Negev.* Manchester: University Press.

—— 1977. The tribe as a unit of subsistence; nomadic pastoralism in the Middle East. *Am. Anthrop.* **79** (2), 343–63.

Moscati, S. 1959. *The Semites in ancient history. An enquiry into the settlement of the Bedouin and their political establishment.* Cardiff: University of Wales Press.

Newcomer, P.J. 1972. The Nuer are Dinka: an essay on origins and environmental determinism. *Man (N.S.)* **7**, 5–11.

—— 1973. The Nuer and the Dinka. *Man (N.S.)* **8**, 109–10.

Ogot, B.A. 1964. Kingship and statelessness among the Nilotes. In *The historian in tropical Africa*, J. Vansina *et al.* (eds.). London: Oxford University Press for Intern. Afr. Inst.

Peters, E. 1967. Some structural aspects of the feud among the camel-herding Bedouin of Cyrenaica. *Africa* **38**, 261–82.

Rey, P.P. 1975. The lineage mode of production. *Critique of Anthropology* **3**, 27–79.

Rubel, P.G. 1969. Herd composition and social structure: on building models of nomadic pastoral society. *Man (N.S.)* **4**, 268–73.

Sahlins, M. 1961. The segmentary lineage: an organization of predatory expansion. *Am. Anthrop.* **63**, 322–345.

Salzman, P.C. 1978. Does complementary opposition exist? *Am. Anthrop.* **80**, 53–70.

Southall, A. 1976. Nuer and Dinka are people. Ecology, ethnicity and logical possibilities. *Man (N.S.)* **11**, 463–91.

Torry, W.I. 1976. Residence rules among the Gabra nomads: some ecological reconsiderations. *Ethnology* **15** (3), 269–86.

CHANGE AND THE BOUNDARIES OF SYSTEMS IN HIGHLAND NEW GUINEA: THE CHIMBU

PAULA BROWN

The Problem: Holism, Equilibrium, and Change

Equilibrium models are now familiar from studies of ecological and social systems, with change treated as a transitional condition between a former and a future equilibrium. A holistic view of ecology, culture and society results in the description of a persistent system of inter-relations in a functional analysis. Yet the study of Chimbu reveals diversity, disturbance, competition, expansion, and other forms of non-equilibrium and change. Change or non-equilibrium may come from many forces or processes: evolutionary adaptation as an internal development, contact and diffusion as external forces later inter-nalised, or conflict provoking revolution, revision or change. Among anthropologists the pre-history or pre-contact period is often thought to be stable while known history can often be turbulent. Prehistoric trends and development are for the most part seen as one-directional and evolutionary; inventions are transmitted by diffusion.

There is a sort of building-block approach to the relations of ecology and society, in which technology and culture are elements: thus environment is a relatively fixed pool of resources, which an ecological system including technology and culture uses or adapts to; and the social system is linked to (or possibly included within) culture. At the same time we often seem to consider environment and resources as stable and unchanging, whereas social and cultural change always occur. Non-equilibrium best describes the Chimbu social and ecological system.

Daryll Forde, who was more concerned with ecological study and influence than most other social anthropologists of his generation, considered the 'body of known and available resources and means of providing and utilising material goods of all kinds' (1970: 19) as exogenous to the social structure. He provides other evidence that he thought of 'external agents' in the same light. But his chief point is that social systems are not autonomous or closed. He argues that the anthropologist should explore the relations between environment, resources, techniques, and social organisation, and also profit from the data of demographers, agronomists and others, yet considering them extrinsic to social structure (1970: 27). Forde introduced the term 'socio-cultural ecology' which 'is concerned with inter-relations between organized human behaviour and the material world' (1970: 26).

In Forde, and in the majority of papers presented in this volume, a holistic approach to social and ecological systems prevails. In this view, those characteristics or aspects sometimes separated as ecological, cultural and social are treated as together forming one system. However, the use of the term 'system' need not require a functional explanatory scheme in which equilibrium is supposed.

An equilibrium model seems to dominate ecological study in biology and anthropology. Cybernetics and feedback mechanisms have been major guiding concepts in ecological study (Margalef 1968; MacArthur 1972). Ecological anthropologists examine the main-tenance and persistence of a local human population (Little and Morren 1976). While the social groups, relationships, or social systems are rarely mentioned or described in any detail, they appear to be an aspect of the human population. The ecosystem is viewed as closed, localised, sustained through feedback mechanisms; homeostasis is the persisting state of the ecosystem. Rappaport, defending his analysis of Tsembaga Maring ritual regulation of the ecological system against criticism by Friedman (1974), states that his analysis is limited to the ecosystem of this local population of 200 people (Rappaport 1977). He considers Friedman's stress on the territorial and inter-tribal expansiveness of some other New Guinea highlands societies, including Chimbu, in their social and ritual exchange festivals, irrelevant to his analysis. Thus, according to Rappaport, while the ecosystem includes culture and society, its dynamics are internal, its boundaries closed.

The dynamics of an ecosystem do include internal cycles and oscillation. Complex repetitive cycles depend on the seasons, harvests, life cycles of animals and land regeneration. Rappaport's Tsembaga Maring analysis is based upon a cycle: ritual pig slaughter follows years of pig herd growth and increasing demands upon human labour and fodder production. In this equilibrium model the processes or cycles are repetitive.

As Collins points out, functional explanation depends upon the isolation of a functional system. But 'to show that an institution functions in a certain way does not by itself constitute an explanation of that institution' (Collins 1965: 276). The problem 'of showing how one functional system has changed into another involves the difficulty of explaining the appearance of new variables and mechanisms' (*ibid*: 280). It 'requires resource to explanatory procedures independent of functional analysis' (*ibid*: 280).

Ecological systems are seen in the long term as evolving and adapting, both endogenous processes. The organism or the community perfects its energy efficiency. Some anthropological ecosystem theorists consider population increase as a stimulus to or cause of technological change and social differentiation. In biological ecology, population numbers of species are elements of the system. Ecological analyses explain the composition of a natural community as an equilibrium among the species. The population of each species is regulated by relations among species in a given environment.

Ecological anthropology is a sort of super-functionalism, not only for cultural continuity, homeostasis of the ecological system, or institutional equilibrium, but for the persistence of the whole ecological system. Clarke characterises the Maring as a palaeotechnic agricultural system with a homogeneous distribution of energy, which can be permanent. This is contrasted with energy distortion and pollution in large energy input systems (1977: 377–80). As Burnham (this volume) points out, Rappaport's 'scheme of analysis is incapable of dealing with social change'. Any persistent system is assumed to be adaptive, so that its functioning is not questioned.

Another focus of controversy concerning Rappaport's homeostatic analysis of Tsembaga Maring is the determining role he assigns to biological factors, especially the natural reproductive and growth cycle of pigs, and the dependent role of Maring social goals. He says, 'it may be doubted if ultimate regulatory authority is ever located in particular personnel' (1977: 172). Both theoretically and in ethnographic reporting this is a genuine point of difference between our studies and Rappaport's; my discussion of pig raising and feast frequency which follows shows that Chimbu leaders conflicted with local opinion about the feast schedule, and that the Naregu people decided when to raise pigs and hold the feast. Hide (personal communication 1974) has studied pig production in a nearby tribe.

The ecological systems analysis uses 'perturbation' and 'disturbance' for events of change or interruption of equilibrium. In studies of natural, non-human communities, examples of disturbances are storms, invasions, winds, floods, and plagues of insects. The mortality, regrowth, and succession of species may not result in return to stability and duplication of the previous ecological system.

Furthermore, it has been observed that species in certain local areas are highly diverse, and in non-equilibrium. In areas of frequent disturbance, and in long-term non-equilibrium, the conventional equilibrium theory can be doubted. Connell (1978) rejects the assumption of equilibrium in all local situations and explains diversity and change of local assemblages as a result of frequent disturbances or gradual climatic changes. There may be an analogy to human social and ecological systems. As applied to the New Guinea highlands, environmental change and cultural introductions, both before and after western contact, may be considered. In Chimbu we have observed changes in land use and land evaluation.

Land and Competition in Chimbu

Our work in Chimbu and the New Guinea highlands, including geography, prehistory, and social anthropology, might be regarded as a convincing demonstration of the interrelations and interdependence of ecological and social systems. Many studies have shown relation-ships in particular local systems, and some broader generalisations seem possible. Applying this approach in the New Guinea highlands, we found correlations of agricultural intensity, land tenure forms, population density and group size (Brown and Podolefsky 1976). We have some prehistoric information and speculation about the develop-ment of New Guinea highland systems (Golson 1977; Sorenson 1976), and from geographic and ethnographic studies we know a good deal of ecological, social, and cultural change in the past 30 years.

Agriculture and pig husbandry, settled community life and, it is assumed, some forms of social groupings and ceremonial activities of New Guinea societies are more than two thousand years old. High-lands culture, as shown by studies in archaeology, history, botany, and ethnography, has changed in the past 300 years since the acquisition of a new food crop. Some time, probably after 1600, sweet potatoes were introduced by indigenous trade and became established as the dominant staple food and pig fodder crop. Then, higher altitude settlement, expanded garden areas, increased population of people and pigs, and larger intergroup exchange ceremonies developed. Discovery of highland peoples in the 1930s found large community groups, patrilineal clans, and competitive ceremonies exchanging food and pigs. It appears to be an expanding ecological and social system: the interdependence of resources, technology and social relations is striking. The greatest development occurs above 2000 metres where only a temporary or very small population could live on foods available before sweet potatoes.

Chimbu intensive agriculture was uniquely adapted to steep hill-sides, permanent occupation by a dispersed but dense population, mobility within and between settlement areas. Stone and wood tools were used for agriculture and construction of houses and fences. Soil fertility was improved with complete clearing, burning of cut vegetation, deep tillage, drainage ditches, mounds for planting sweet potato vines, gradual harvesting, preparation of ditches and mounds for replanting, cultivated casuarina trees, fallow periods, and the integration of pig keeping with forest fallow, cultivation, and settle-ment areas. Chimbu agriculture is intensive and well adapted to the steep terrain and crop requirements. In the areas of highest popu-lation density, permanent fenced fields are maintained and re-cultivated with only a brief fallow period.

Sweet potatoes are the main food crop for human and pig consump-tion. Large pigs are required for a feast. Land needs oscillate with the ceremonial cycle; larger areas are planted with sweet potatoes to fatten pigs and to provide for the increased needs of hosts and guests before a feast. Since a Chimbu pig feast requires production and cooperation of a whole tribe, and sets of tribes coordinate their feasts, thousands of families are involved in production of pigs, building ceremonial villages, obtaining decorative feathers, attending ceremonies, dis-

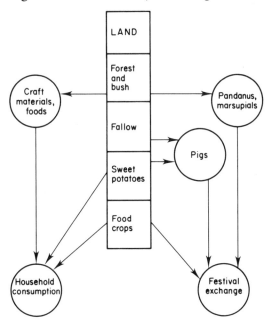

Fig. 1. The Chimbu use of land resources for household consumption and festival exchange.

tributing and consuming meat. No local group or individual is isolated from this large competitive exchange system: the system is unbounded and expanding.

The life cycles of people, pigs, trees, and crops made land and resource utilisation complex. In a careful analysis of ten years' data on Chimbu, H.C. Brookfield shows how the complex interrelation of cultivation cycle and pig cycle:

> entails a complex web of spatial arrangements designed to accommodate the strains of a system which varies its demands on resources through a factor of more than two, each few years. No interpretation of man/land balance or of spatial organization of activity, can be valid unless it takes account of this repeated expansion and contraction (1973: 159).

In ecological and social systems a great number of forces are inter-related: regularity is normal in annual seasons, life cycles of plants and animals, the use of land by swidden cultivators, as well as in family, domestic, and ceremonial cycles. The continuity of environment and social institutions supports such regularities. Yet the interrelations of these produce a complex pattern, and disturbances by catastrophic natural forces (such as hurricanes and epidemics), social conflict and warfare, inventions and diffusions, all may have a significant impact upon a social and ecological system.

The area now comprising the northern part of Chimbu Province is the largest area of dense settlement in New Guinea. The overall population density on land below 2400 metres is 84 persons per km^2 (Howlett 1977); local densities are often much higher. Mountain conditions are healthy and favourable to population increase. In steep mountain slopes, highly intensive techniques of cultivation and pig raising are practised. Chimbu tribes are competitive: disputes and wars displace families, communities reorganise and reallocate land, ceremonial exchange builds tribal reputation. Pig production requires fencing, forage areas and fodder crops. which strain land resources. The land and food needs of pigs and people were the focus of tensions and land conflicts. Huge ceremonial villages accommodating several thousand people were built for intertribal ceremonies. The Chimbu people were expanding in all directions from their traditional origin-place in the Chimbu River Valley. Large populations and big tribal groups developed in adjacent valley systems to the east, south and west, throughout the altitude zone between 1,500 and 2,400 metres. By the time of discovery, the frontiers of Chimbu expansion were only at places where other peoples were previously established, where mountains were too high for cultivation, or at distinctively different topography in the southern part of the province. Migrant groups outside the Chimbu Valley had to adapt to different land and agricultural conditions and some other conditions such as malaria at lower altitudes.

In 1930 competition was severe, leading to migration and expansion. Especially on the higher altitude slopes, free of malaria, a permanently settled, dense, but dispersed population practised intensive agriculture. Local and patrilineal kin groups were allied into large exogamous clans and these into tribal groups of thousands of people, in fluctuating conflict and alliances. Between tribal concentrations were shifting buffer zones but valued land was continuously guarded and used in individual tenure. War, migration, and conquest reallocated land. Interpersonal and intergroup festivities exchanged pigs, shells, axes, feathers, and other valuables. Big men directed ceremony and conflict management in tribes. The major political relations and ceremonial exchange integrated the productive capacities and mobilised fighting forces of hundreds of men and their families.

What might have happened if discovery and the Australian colonial programme had not intervened?[1] I believe there might have been expansion and adaptation to other climatic zones in southern Chimbu, increasing competition, and further outlying settlement in the Wahgi Valley region. The pressure for ever-increasing pig prestation may reach a limit, if not diverted into other goods or produce (Strathern 1976). While the Chimbu Valley habitat is favourable to intensive agriculture, further population increase has severely strained the technology and resources. Beyond the level of technology reached, some further improvement in agricultural techniques may have developed. The garden damage and food demands of large pigs would, I believe, become intolerable and some changes in the husbandry techniques or feast frequency might develop so that pigs would not so often break into gardens.

Interrelations of Land, Subsistence, Cash Crops and Ceremony in the Modern Economy

The arrival in 1933 of an Australian exploratory party, missionaries, and shortly thereafter establishment of administrative posts had its earliest impact in stopping fights, fixing group boundaries, and regularising group membership with census books. The Australian control programme established a new legal system, local political representation, and limited territorial expansion. Health, education, local government, and economic development followed. For 20 years the Chimbu were only rarely visited by outsiders; after 1951 labour migration and local development projects brought many changes which affected the ecological and social system. Under Australian administration, the development programme included health and educational services, introduced some cash crops (although not for the upper Chimbu Valley people) and allowed population resettlement in land with cash crop potential by organised schemes and individual or family migration. The population has continued to grow.

Local changes include changing land use, new economic activities, expanding and diversifying exchange activities. The traditional clan and tribal groups are now somewhat re-assorted and combined in larger local government units. While the physical environment is hardly changed, communications, transport, and trade have expanded local knowledge and penetrated the former isolation of Chimbu.

The great value placed on interpersonal and intergroup competitive exchange by Chimbus, which has been observed in the past, continues to direct their food production, pig and cattle raising, cash crop sales, and savings activities. People are more mobile, and their exchange relationships are wider afield since the development of road transport, and long-distance affinal and friendship ties. Resettlement and cash crop production are both opportunities for individual achievement and may be incorporated in exchange. Ties to one's own group and land, and patterns of exchange, debt and credit perpetuate the inter-connections. Change and substitution of trade goods, purchased food, cattle, and money have not substantially affected the integrative and group prestige significance of exchange. Scale, in quantity and variety of goods, and range of social ties, have increased.

Stone and wood tools have been replaced with metal, especially steel axes, shovels, and knives. These are more efficient and save considerable time in land clearing, garden preparation, planting, fencing, housebuilding and other agricultural and domestic labour. Some construction materials, personal clothing, and household goods are now purchased. The domestic economy is no longer wholly self-sufficient, but the majority of foods and materials are still produced locally.

Chimbus, as many other Melanesians, were attracted by western goods and activities from their first view of them. I first saw this enthusiasm as a hope for replacement of traditional ways with western ones: speeches and plans were to imitate western ways. The leaders of the 1950s and early 1960s had great hopes for development of individual family business, communal or cooperative ownership of trucks, coffee processing plants, and shops, with improved living conditions and a rising participation in the colonial economy. For a time in the 1960s pig ceremonies and funeral exchanges were almost eliminated, at least in the tribe of the most forceful progressive leader. Efforts were re-directed to coffee growing, cash crop gardens, and local government council projects. A number of developments might have been expected from this beginning. However, these alternatives were not followed. Chimbu has not seen a development of independent peasantry with family goals. Commercialisation, individualisation, land consolidation, and specialisation are little developed. Such developments as economic specialisation and shifting to cash crop production only have been seen as a possible alternative among some neighbouring eastern and western highlands peoples (Howlett 1973; Finney 1973).

It is recognised that pig feasts make many demands on land and labour, which may be incompatible with the drive for modernisation and cash production. Leaders attempted to put an end to these ceremonies so that coffee and other cash crops, and business could predominate. In the high density areas, as land was permanently planted in coffee, the amount of sweet potatoes required to feed pigs in the year or two preceding a feast could not be produced; the two land uses are competitive (Brookfield 1968: 973). Indeed, the Naregu pig feast was held off for many years. However, this could not control all pig raising efforts; some pigs are required for smaller exchanges such as marriage. Most Chimbu tribes continued to raise pigs for feasts and coffee for market, perhaps in smaller amounts than in the eastern and western highland regions where population density was less and commercial development greater. The popular demand for a pig feast in Naregu tribe of Chimbu grew after the death of the most forceful pro-business leader, and it was held in 1972. Since then still other factors and interests have affected feast and cash crop activities. These observations of the role of leaders and the participation of family heads in feast preparations clearly demonstrate the choices and planning of individual Chimbus in group activities.

There is some change in the division of labour and allocation of time as well. Cash crop production and sale, church services, schools, travel, and visiting between home, town, clinic, market, and mission all now vary the daily and weekly round. News of meetings and activities is disseminated by radio; local leadership has declined in importance. Provincial government activities may take their place.

Although the Chimbus eagerly sought western goods, and applied themselves to coffee growing, marketing foodstuffs, and any other sources of money available such as migration for labour and tourist entertainments, their land resources and population density severely limit development. While about 0.18 ha/capita is in food crops, coffee land reached 0.06 ha/capita at Mintima in Naregu tribe (Brookfield 1973).

Livestock have always been important to Chimbu, and this is reflected in recent development projects and government-supported loans. There are two types: intensive pig raising in piggeries, with hand feeding, which requires cultivation of sweet potato fodder and supplements; and cattle projects which fence grassland and often require additional fodder. Both pigs and cattle are used mainly in prestations rather than as commercial investments, although this was not the intention of development project programmes.

The present land evaluation is quite different from that of 1933, despite the continued domination of the land use with agriculture and the scattered homestead settlement form. There has been some

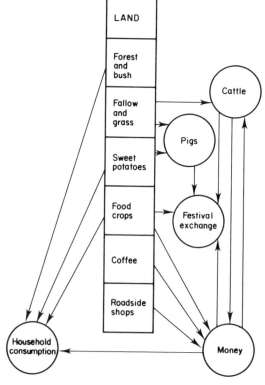

Fig. 2. Chimbu land and resource allocation today.

change in the evaluation of resources. Although in short supply, subsistence land has been available to all, but land for new activities is not equally available. Land suitable for coffee is unevenly distributed; coffee planting ranges from a high of 95 trees per capita in the far east of the province to 4 per capita in the densely settled high altitude Mitnande. But since the 1950s and 1960s planting developments, there has been little new planting, and many young people have no coffee trees of their own (Howlett 1977).

Aside from agricultural land, the most important resources are sites for houses and commercial enterprises. Here the former value was location in relation to the interests of the people and security from attack. Now, the roads are probably the most important element in site selection. Those groups with highway access have more favourable commercial opportunities, get higher prices for their coffee, and have more contacts with outsiders. A roadside site is necessary for shops.

The Chimbu physical environment has changed little, but the Chimbu now use it differently and expect different returns from it. While it was formerly valued for its natural resources, wild plants, and potential for food crops and pig forage, cash crop interests create a different evaluation. Coffee has been the main crop developed, and its limitation to a lower altitude than has been used to grow sweet potatoes has created some discrepancies. Coffee cannot be grown in the upper Chimbu Valley, which is the area of greatest population and the traditional Chimbu homeland. Land in the upper valley has no cash crop economic potential at present and is now inadequate for subsistence to the population. Pyrethrum has not succeeded (Howlett 1977). Furthermore, the area is steep, the road dangerous, and the entire area has been bypassed by highway developments. The dense Marigl Valley, Gumine sub-province area, has somewhat different problems: later contact, isolation, but possible expansion of settlement to the south.

At Mintima, where the highway and suitable coffee land stimulated cash cropping, there was some substitution of cash crop production for subsistence and feast foods. Those Chimbus whose land is too high for coffee or remote from roads have no such opportunities. They have exported labour and applied for resettlement in other areas – the people have diversified their activities and depend in some part on earnings from outside the area (Howlett 1977). This is true also of the coffee-growing areas, as here men have chosen voluntary resettlement in highland areas more accessible from their homelands. Some maintain close ties, and travel frequently between old home and new land areas.

These new sources of money, developed mostly in the past ten years, and quite dramatically when coffee prices increased in 1976, brought unprecedented spending power to an area of poor subsistence farmers. Throughout the period I have observed, money, as cash and as purchased goods, has been placed in the exchange economy. Traditional marriage payments have been supplanted by cash and new valuables, and greatly inflated. The large food exchanges (*mogena biri*), which take the form of a series of person-to-person prestations assembled in a sequence from numerous groups to the donor group, and a final grand collective prestation of individual gifts by members of the donor tribe to individuals in the recipient tribe, are now made up of both cultivated and purchased foods, beer, cattle, and money. Traditionally, these food prestations interchanged wild and cultivated produce of different ecological zones; now these are of minor importance, although cattle may perpetuate this pattern since they are mostly limited to the lower altitude open valley region. Money has greatly inflated the payments. The expansion is similar to that described by Strathern for the Melpa (1976).

A result of the particular crops and projects introduced is to benefit those Chimbu groups who were the evicted, migrant peoples — pushed out of the high narrow steep valley that was the Chimbu homeland and into the open broad valley where density remains lower. They had poorer land for subsistence and were somewhat exposed to malaria; now they are better placed for communications, coffee, and cattle. There have been a few opportunities for advanced or technical training and commercial development in Chimbu, and these have been somewhat selectively available to those close to mission and government schools. These, too, were first placed on land that was regarded as undesirable by local Chimbus.

Some of the land that was not closely occupied was a buffer-zone between enemy tribes and kept the enemy groups apart. But war zones and no-man's-land zones had been shifting and available for reoccupation later. The increase of population throughout, cessation of fighting, fixing of boundaries, and allocation of some land to government, commerce and other uses has severely restricted traditional modes of adjustment of land needs. Both flexibility and the possibility of gaining new land by migration, encroachment or conquest were stopped by colonial policy, or its local interpretation. For about 30 years, from 1939–1969, Chimbus were afraid of Australian colonial officials and expected land disputes to be settled in courts. More recently they have revived warfare and retaliation in land and other disputes. This has increased since independence in 1975. Compensation payments following accidental death involve very large payments, and threats or actual physical attacks are frequent during negotiations. In their reactions, threats, and impulse to physical retaliation the Chimbu often seem to hold a conflict theory of society: only coercion will control them, they say.

The wealth-display-pride-prestige of distributions demonstrates group and big man success in production. This is now most emphasised in local exchanges. The exchange system before 1933 served to distribute scarce goods between areas, and the interpersonal relations honoured in exchange could be activated for help or sanctuary at times of need when a family suffered losses. Ties to distant kin and affines extend the scale of visits and increase the range of goods distributed; the visits and invitations are themselves a source of prestige. These distant connections, especially with prestigeful outsiders who are educated and employed in towns, are welcomed. Many originate in liaisons between Chimbu women and the outsiders; their fathers and brothers may not receive the quantity of marriage payments that they would obtain from a Chimbu, but they expect urban hospitality.

Many new developments indicate the beginnings of social stratification, as well as diversification and specialisation. These naturally involve ecological considerations, access to and control of newly valued resources, knowledge to exploit resources, and command over the labour of others. It is especially in this field that the crucial role of technology in mediating between the environment and social system can be appreciated. Successful entrepreneurs gain advantages through knowledge. The first leaders in cash cropping and trade acquired land for agriculture or commerce which at the time was available because it was low-lying, exposed battleground, or for some reason avoided by local people.

In all of these developments the Chimbu have been disadvantaged by their former success and growth: dense population in steep valley slopes, large clans and tribes have not attracted commercial ventures. The peoples in provinces to the east and west have witnessed Australian investment for development, whereas in Chimbu land has not been available for settlers or model development projects. The most energetic Chimbus have left the province for settlement in unoccupied land, easily adjusting to the physical conditions and developing independent enterprises. This individual peasant holding pattern may become a new ideal, but it cannot be achieved in the traditional Chimbu homeland.

Non-equilibrium

Studies in the New Guinea highlands have stressed the role of the sweet potato in the process of adjustment and expansion of the pre-contact period (Watson 1977). Local systems may have unique features, but a general similarity is undeniable, and some interchange of technology and belief must have taken place. The highland peoples and groups are open, engaged in trade and exchange over large areas which introduced products and practices from afar. Imported goods were valued, if not always essential to survival. A homeostatic equilibrium model cannot adequately describe Chimbu. Chimbu was both open and expanding, competitive within and pioneering or conquering on its borders. I must emphasise the expansion and competition in Chimbu, while a less volatile characterisation might suit some other groups. The homeostatic model has not been seriously questioned in New Guinea studies (Bayliss-Smith and Feachem 1977) even though Clarke (1971), Waddell (1972) and others have discussed change and diversity in ecological systems.

An equilibrium model seems inapplicable to post-contact Chimbu. It is not that the pre-contact values and practices are gone, to be replaced with others, thus resulting in another social and ecological system. Rather, many if not all of the earlier practices and values persist along with alternative, often contradictory ones. The Chimbu of today want both coffee crops and feasts, money for consumption and exchange, subsistence foods and purchased delicacies, home settlement and town jobs, peace and war. This is not a regional or generation gap so much as a new diversity of interests and re-evaluation of resources and opportunities. A great diversity of career goals was mentioned by high school students in reply to a questionnaire in 1976.

Exogenous factors, colonial dependence, entry into the world economy, and self-government are only in part responsible for these discrepancies and confusions. However, modernisation is not responsible for the non-equilibrium of the Chimbu social and ecological system in earlier times.

The long term study of the Chimbu social and ecological system as it has existed throughout the known period has shown many continuities and interrelated forces. New elements in the economy, government, and society have been incorporated to make the modern system. However, neither the past nor the present has been, or seems to be moving toward, equilibrium. The examination suggests that the closest analogies in other sciences are non-equilibrium of ecological systems and conflict in society. Figure 3 depicts a number

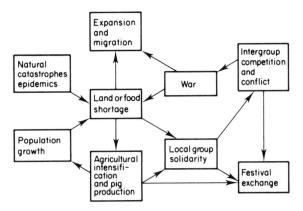

Fig. 3. The Chimbu: a non-equilibrium system.

of interacting forces of the social and ecological system. Most of these are interrelated. However, natural catastrophies and epidemics are shown as of external origin; expansion and migration go beyond the area of occupation.

Today, this connects Chimbu with Papua New Guinea and beyond. Population is dependent upon food and resources. Local group solidarity and intergroup competition and conflict are general conditions of New Guinea highlands peoples, found among peoples with land surplus as well as land shortage. Thus I show land and food shortage as a result of war and festival exchange as well as population growth. More direct connections can be found in the population growth − agricultural intensification cycle and in the production − exchange cycle. These together have made the Chimbu expansive. The favourable environment of the Chimbu Valley was the area of greatest densities, largest festivals, most intensive agriculture, and emigration. It has little potential for cash-productive enterprises but continues as the homeland of emigrants.

In social and natural science, recent observations and theoretical developments have questioned the functional equilibrium assumptions of the past. A simplified homeostatic model does not adequately represent reality in all situations. Among ecologists, Slobodkin, who pays particular attention to man's special characteristics, notes that man may change his interpretations when no environmental change has occurred (1974, 1978). Connell (1978) cites many studies of diversity and non-equilibrium in tropical rain forests, and Porter shows the competitive and diverse conditions of coral reefs (1974). Homeostasis and adaptive equilibrium are not found in all human communities. Diversity, as in the subsistence of system of Gadio (Dornstreigh 1977), competition and expansiveness are elements of a dynamic non-equilibrium such as exists in Chimbu.

Footnotes

1. I am indebted to J.A. Barnes for pointing out a possible misinterpretation of this discussion.

   It is not here suggested that the Chimbus had reached a crisis at the time of discovery, but that strains on resources, competition, agricultural intensification and expansion into surrounding areas had been proceeding for a century or more. While technological adaptation and cultural change might take place in these circumstances, the contact situation and subsequent changes have created new problems and solutions, including improved health and longevity, temporary cessation of warfare, and opportunities for long distant migration, resettlement and new sources of income.

References

Bayliss-Smith, T. and R.G. Feachem (eds.) 1977. *Subsistence and survival.* London: Academic Press.

Brookfield, H.C. 1968. The money that grows on trees. *Australian Geog. Studies* **6**, 97–119.

—— 1973. Full circle in Chimbu: a study of trends and cycles. In *The Pacific in transition*, H. Brookfield (ed.). Australian National Univ. Press.

Brown, P. and A. Podolefsky. 1976. Population density, agricultural intensity, land tenure, and group size in the New Guinea highlands. *Ethnology* **15**, 211–38.

Clarke, W.G. 1971. *Place and people.* Berkeley: Univ. of Calif. Press.

—— 1977. The structure of permanence: the relevance of self-subsistence communities for world ecosystem management. In *Subsistence and survival*, T.P. Bayliss-Smith and R.G. Feachem (eds.). London: Academic Press.

Connell, J.H. 1978. High diversity in tropical rain forests and coral reefs. *Science* **199**, 1302–10.

Collins, P.W. 1965. Functional analysis in the symposium 'Man, culture, and animals'. In *Man, culture and animals*, A. Leeds and A.P. Vayda (eds.). Publ. 78, Am. Assn. Adv. Sci. Washington, D.C.

Dornstreich, M.D. 1977. The ecological description and analysis of tropical subsistence patterns: an example from New Guinea. In *Subsistence and survival*, T. Bayliss-Smith and R.G. Feachem (eds.). London: Academic Press.

Friedman, J. 1974. Marxism, structuralism and vulgar materialism. *Man (N.S.)* **9**, 444–69.

Forde, D. 1970. Ecology and social structure. *Proceedings of the Royal Anthropological Institute for 1970*, 15–29.

Golson, J. 1977. No room at the top: agricultural intensification in the New Guinea highlands. In *Sunda and sahul*, J. Allen, J. Golson and R. Jones (eds.). London: Academic Press.

Howlett, D. 1973. Terminal development: from tribalism to peasantry. In *The Pacific in transition*, H. Brookfield (ed.). Canberra: Australian Natl. Univ. Press.

Howlett, D. *et al.* 1976. *Chimbu: issues in development.* Development Studies Centre Monograph No. 4. Canberra: Australian Natl. Univ. Press.

Little, M.A. and Morren, G.E.B. 1976. *Ecology, energetics, and human variability.* Dubuque: William Brown.

MacArthur, R.H. 1972. *Geographical ecology: patterns in the distribution of species.* New York: Harper & Row.

Margalef, R. 1968. *Perspectives in ecological theory.* Chicago: University of Chicago Press.

Porter, J.W. 1974. Community structure of coral reefs on opposite sides of the Isthmus of Panama. *Science* **186**, 543–5.

Rappaport, R. 1968. *Pigs for the ancestors.* New Haven: Yale Univ. Press.

—— 1977. Ecology, adaptation and the ills of functionalism (being, among other things, a response to Jonathan Friedman). *Michigan Discussions in Anthropology* **2**, 138–90.

Slobodkin, L. 1974. The peculiar evolutionary strategy of man. Lecture presented to the Boston Philosophy of Science Colloquium, Octobter 8, 1974.

—— 1978. Is history a consequence of evolution? In *Perspectives in Ethology*, P.P.G. Bateson and P.H. Klopfer (eds.), **3**, 233–55.

Sorenson, E.R. 1976. *The edge of the forest.* Washington, D.C.: Smithsonian Institute.

Strathern, A. 1976. Transactional continuity in Mount Hagen. In *Transaction and meaning*, B. Kapferer (ed.). (A.S.A. Essays in Social Anthropology I). Philadelphia: ISHI.

Waddell, E. 1972. *The mound builders.* Seattle: University of Washington Press.

Watson, J.B. 1977. Pigs, fodder and the jones effect in post-ipomoean New Guinea. *Ethnology* **16**, 37–70.

HEGELIAN ECOLOGY:
BETWEEN ROUSSEAU AND THE WORLD SPIRIT

JONATHAN FRIEDMAN

In a very elaborate and useful reply to a critique of cultural ecology that I wrote some years ago (Friedman 1974), Roy Rappaport has made an explicit attempt to clear up some of the difficulties of the ecological approach as well as to provide a rather full-blown general model of the relation between societies and nature in history. The issues raised are somewhat peripheral to my present interests, but the rather polemical response to my admittedly over-polemical attack reveals, as far as I can see, that the general approach has not changed. Instead, it is now clearly linked to what appears to be a deeper ideological framework. The following is meant to be a partial reply to Rappaport. I do not intend to take up the specific criticisms and contributions of his article (Rappaport 1977a), many of which I agree with totally. Rather, I would like to try to get at the underlying ideology that pervades the *a priori* assumptions that I criticised earlier and which today — in a period of major crisis — is even more evident and, I think, more dangerous than previously.

In order to make clear the ideological principles behind the theoretical model I shall proceed from the clearly ethnographic and analytical discussion to the moral-political statements that are apparently of a higher 'logical type', as it is said, and which order the lower order explanatory models.

The Cybernetic Savage

As I presented my case, the cultural ecological model was a model of functional causality where institutions were rational creations of the ecosystem whose purpose was to fulfil specific functions with respect to the maintenance of populations in their environmental niches. The weaker form of the functionalist argument was merely a pseudo-descriptive statement of systemic relations said to exist between institutional complexes and environmental variables. I would claim today that my critique might easily be extended to many of the marxist analyses, especially to Godelier's discussion of the Mbuti (1973) and of the Australian Aborigines (1975).

The model which I advocated at that time was one in which adaptation was not an active process but a temporary state of affairs: one determined not by a specific relation between structures and their conditions of reproduction but by the simple existence of degrees of freedom within larger systems. As such, local structures survive as long as essential limits are not transgressed, but there is no *principle of non-transgression*. This is especially true of social systems that are not planned in order to optimise their possibilities of survival but have entirely different and more destructive goals. The proposed model was based on contradictions, incompatibilities and transformations, where explanations were necessarily historical-evolutionary rather than function in character. Here technological constraints play the role of limiting factors or perhaps buffers (Waddington 1968), when seen as temporal functions, that act as bounded pathways in morphogenesis. The fact of compatibility between structures and their external conditions, however, can in no case constitute an adequate explanation for the existence of those structures.

Serious objections can and have been raised with regard to the model that I proposed. These primarily concern the restriction of the formal model to the single society and its techno-environment, thereby implicitly assuming that local societies are closed reproductive units (Ekholm 1975, 1976; Friedman 1975a). Where local societies are not reproductive totalities they are no longer sufficient units of explanation. This, however, does not affect my argument against the functionalism that continues to underly cultural ecology and its more elaborate anti-materialist variants.

Adaptation, for Rappaport, is a process and not a mere state of affairs. This can be very tricky, especially as adaptive processes can be defined as those that lead to survival, either by homeostasis or by transformation (Rappaport 1977a: 168), i.e. processes defined in terms of their results, a state of affairs, This has much to do with the difference between teleological and teleonomic processes. The first is organised action with respect to a specific goal where the goal is itself an integral and explicit part of the activity. The second refers to the programme-like nature of an activity where, however, the goal of the activity is not included in the programme, being on the contrary a predictable result of that programme. Epigenesis is thus teleonomic but not teleologic insofar as the fully formed organism, while predictable from the genotype, is not contained as information within the genetic programme. The organism is, rather, the result of chreodic processes (Waddington) of which the genetic programme is the basic element. There is a crucial difference between jumping off a cliff and walking, in a straight line, off the end of a cliff in the dark. Both actions are characterised by intentionality, but only in the first case does the intention correspond to the result. It seems to me that Rappaport's and even more so Bateson's approach tends to purposely obscure this distinction. It is only in this way that ecosystems can be conceived as subjects that,

> ... not only transform themselves in response to changes in external conditions, sometimes by replacing all of their constituent species populations with populations of other species, but also through the mutual adaptation of their constituent species to each other (Rappaport 1977a: 149).

Perhaps I have misunderstood biological reasoning here, but it seems to me that the properties attributed to processes/operations/ mechanisms called adaptive are not inherent properties of the operations themselves but refer to one aspect of the result of those activities, namely continued survival. I cannot see how 'survival' can be attributed as a function to a process where there is no clear teleological organisation of that process. If, on the other hand, 'adaptation' is meant only as a classificatory concept that groups processes with respect to their outcomes, then we are indeed dealing with a very low level concept − i.e. all those processes that do not, for whatever reason, disturb the systems of which they are a part. Thus, the fact that people die is adaptively related to the long term maintenance of human population below certain crucial limits (e.g. carrying capacity). The same is true of all other life-destroying activities, warfare, murder (infanticide). All of these can be added to conscious attempts to limit birth rates. To group so many different phenomena under the same heading is a rather trivial exercise.

If adaptation is not a property inherent in the processes to which it is ascribed, it becomes rather uninteresting. If, as I think Rappaport means it to be, adaptiveness is inherent by definition, then I would claim that it is falsely conceived insofar as it conflates external effect with internal process.

It appears to me now after reading Rappaport's critique that the question of teleology is a major point of argument, so I will try to elaborate further here by returning to the old discussion of ritual as a homeostat.

Rappaport, in his criticism of my assertion that Maring ritual cycles do not function as a homeostat with respect to environmental limits, claims that,

> Friedman misunderstands negative feedback. Negative feedback is simply
> a process in which deviations from reference values initiate operations
> tending to return the state of the deviating variable to its reference
> value. The relationship of reference values to goal ranges (what Friedman
> means by limits, i.e. the possible range of viable or homeostatic states) is
> a separate question . . . (Rappaport 1977a: 158).

In my discussion (1974), I make use of the thermostat as a model
of a homeostat in which the trigger is a dependent function of the
limits below (or above) which controlled variables are to be maintained.
What this means is that there must be a specific objective relation
between the limit temperature in a room and the trigger. All this
corresponds to Rappaport's own definition, where reference values
(not goal ranges) are equivalent to my 'limits'. My argument was that
the limits of environmental maintenance are included in the regulatory
model even though they do not function as reference values for the
ritual homeostat.

(a) I say that my 'limits' are equivalent to Rappaport's reference
values and not his goal ranges because, as I have defined it, the
reference values must be dependent functions of the goal ranges if
the latter are to be included in the cybernetic description. This is to
say that goal ranges must be functionally 'represented by', i.e.
equivalent to the reference values that are actually present in the
regulatory process. In a thermostat, the goal range and reference
values are identical. It is here that our misunderstanding and/or
disagreement lies.

If Rappaport insists that environmental limits are goal ranges and
not reference values and does not demonstrate a functional relation
between the two, then such limits are not necessarily included in
the regulatory process itself. As such, the central argument, the ritual
regulation of the environment, falls to the ground. There is no
homeostatic regulation of the environment but rather the maintenance
of certain environmental variables as a non-intentional result (of which
there must be an indefinitely large number) of the ritual cycle.

(b) A more general point: a thermostat is a mechanism that must
be set by a human regulator. The act of controlling temperature is in
this way teleological. I cannot see why it is anything but misleading
to assume that the Maring set their ritual regulator or that the
institutional structure of the ritual cycle is like a homeostat. I see no
evidence of that teleological quality that is crucial for the operation
of a thermostat. My point here is that the notion of homeostat in the
case of the Maring does not help to account for what actually occurs

but instead adds what appears to be a lot of superfluous information. This has to do very much with the whole starting point of the analysis. It is because the cybernetic model is used — because regulation is what the Maring system must be about — that there is no analysis of the 'reference values' themselves, of the social structure and concrete strategies that determine the functioning and dynamic of the system. It is simply assumed that the primary aim of the Maring is self-perpetuation and that all the rest is secondary or even ideological.

(c) The timing of Maring pig slaughters is determined by pressures within the functioning system. The question here is whether the conflicts are a signal carrying information telling the population 'o.k. do it now', or whether there is a real objective conflict between the drive by men to accumulate pigs and the level of exploitation of women and/or a more general level of social conflicts. I would argue that the interpretation of such situations as information bearers is entirely superfluous. There is a world of difference between the conception of the ritual cycle as an active device intervening in the process of accumulation of pigs in order to regulate specific variables and the view that the ritual cycle functions passively, triggered as a result of the social impossibility of continued accumulation.

The social form of the Maring economy links feasting to a relation between the living and their ancestors, a very common feature in a great many tribal societies that do not have the same kinds of problems as the Maring. The fact that even in New Guinea a similar ritual structure is linked to widely different production levels and intensities might induce us to consider a less regulatory approach to the problem.

The killing of pigs is the immediate regulator of the system in the sense that the pig population is quite simply reduced (regulation= reduction). The content of the ritual in which it occurs does not appear to be symbolically linked to the ecological result but is a much more general structure that perhaps connects the prestige of the pig feasting group to its ability to distribute pork to its allies — a relation in which the prestige of the group is expressed by its relation to its ancestors. If this is the case, and it seems to me that this issue should have been analysed much further, then the content of the ritual structure is unrelated to the ecological outcome of that ritual. Thus it might again be argued that the ritual is not a homeostat. It is the conflicts in Maring society that lead to an action that main-tains in fact, if not in purpose, certain environmental conditions. The accumulation of pigs generates, after a period, conflicts of such an order that the process must be halted. The feast can then be given because no more pigs can be accumulated, a feast whose function is related to the establishment and maintenance of alliances (regional) and the gaining of prestige and political followers (two thirds of the pork produced for a festival is given to other local groups: Rappaport 1968: 214). As the concrete goals of the operators in the system are the accumulation of prestige and control over others' labour and military-political allegiance, men would certainly go on accumulating in order to give bigger feasts if not for the increasing strains in the system. That this process can be analysed as an organised regulatory mechanism within which the conflict ceiling must ultimately be seen as a thermostat setting seems to me to be a clear case of over-interpretation. It brings Maring society to the verge of being a Rube Goldberg machine.

(d) Put briefly, then, I find that the cybernetic analysis of Maring society is unnecessary to account for phenomena that are due to more mundane mechanisms. The fact that societies do not always destroy their own conditions of existence need not be explained as a result of human planning. The fact that the slaughter of pigs is ritualised has, I think, to do with the nature of Maring as well as other tribal social relations and not with a regulatory function. This is not to deny the possibility of describing, at least, the relation between pig killing, conflict levels and labour intensity as a cybernetic one. As for the larger environment/society complex, it is of course possible to use a cybernetic description for any system that does not destroy itself (in the period of observation) by interpreting *de facto* survival as self-regulation.

A thermostat is in a direct causal relation with room temperature. This is not the case for the so-called ritual homeostat. The real trigger mechanism is the tolerance limit of people involved and not the ritual cycle. It is rather mystical to assume that the people need the ritual to tell them that the limit has been reached. An alternative approach is one that considers the actual strategies and structures in the system, one which allows us perhaps to understand the relation between groups like the Maring and Highlanders that, with much of the same ritual structure, have many more pigs (especially as a function of time, see Hide 1974), reduce conflicts by a stricter control over herds, work the land much more intensively — leading perhaps to much higher population densities, overpopulation problems, more serious warfare, ecological degradation, even technological change (see Brookfield 1973a, b).

Systems need not be cybernetic in order to be systems. Cybernetic systems are specific in that they are managed and regulated hierarchically. The systems to which I refer, and which I think are the normal case for human social systems that are not self-conscious entities, are those in which there are numerous processes and tendencies that are basically contradictory to one another. There are, of course, numerous limits that are never exceeded, but for reasons that have nothing to do with any regulatory procedure with respect to those limits.

Ideological Foundations: From Savage Mind to Mindless Savage

Moving from the concrete analysis of a specific society, it is possible to link the cybernetic approach to more abstract assumptions that are ultimately rooted in the clearly ideological matrix of 'Mind' (Bateson 1972) — or managerial pantheism.

The argument against the cybernetic view of Maring ritual is based on what I claim to be *a priori* supposed regulatory operations which I consider unnecessary to explain the 'maintenance' of environmental conditions.

The cybernetic argument is closely linked to the notion of adaptation which is envisaged as an active process of self-regulation that, as we saw previously, includes both homeostasis and transformation. In this view the ecosystem, species populations etc. become subjects that regulate their self perpetuation through their own transformation.

What, we may ask, is an entity that preserves itself by its own self-transformation? Where is the 'self' in all this, if it is not just an empty subject? Rappaport asserts that self-maintenance is the same kind of phenomenon as evolution which is described as 'self-organisation'. The process of selection — implying elimination as well as the survival of some variants — is reduced to the organised activity of a larger entity, the ecosystem, that continually replaces and transforms its constituents in order to maintain itself. Where and who is this 'self' that is apparently so active? Are we not dangerously close to the Hegelian 'identity of contraries' — the species that becomes another species in order to maintain its self? Is not the notion of ecosystem as Subject strangely similar to Hegel's 'World Spirit'? Is it necessary to impute such anthropomorphic qualities to Nature in order to account for transformations? I think not — unless one finds it necessary, for some reason, to have the universe well under control.

Self-organizing (or 'structural' or 'evolutionary') changes in components or aspects of systems are 'functions' in the light of the self-regulating processes of the larger systems of which they are parts (Rappaport 1977a: 170–71).

It is one thing to say that the transformation of local systems is functionally related (in the mathematical sense) to their place in larger systems and an entirely different thing to suppose that their transformation is a function (dependent) of a larger regulatory mechanism. It is the difference between teleology and teleonomy, the difference between system and MIND.

It is further assumed that all living systems are by definition adaptive and that the specific structure which makes them so is cybernetic and hierarchical. Orderly adaptive systems are those in which, 'subsystems and regulators should be hierarchical along a number of dimensions' (*ibid*: 174). These dimensions include:

(a) specificity of goals
(b) response time
(c) reversibility.

The regulative hierarchy referred to here is well known in the literature (Rappaport 1971; 1977b; Flannery 1972) and is derived in part from the work of Bateson (1973) on logical typing and adaptation. Control hierarchies are hierarchies of increasingly general rules of the type:

The control structure here runs from top to bottom. It is further assumed by Rappaport that higher order regulators should, 'include within their repertories programs for changing structurally, or even replacing, lower order regulators'(Rappaport 1977a: 174).

In such systems, the more abstract regulates the more concrete, or at least ought to under normal conditions. Thus, in social systems, religion occupies a higher place in the control hierarchy than the 'economy'. Where it does not do so we are clearly faced with a case of maladaptation – an 'error of logical typing' (*ibid*: 182). Rappaport is extremely interested in religion as such and stresses its inherently adaptive nature.

> Being devoid of social specificity they [propositions about God] are well suited to be associated with the general goals of societies, namely their own perpetuation, for they can sanctify changing social arrangements while they themselves, remaining unchanged, provide continuity and

meaning through those changes (*ibid.*: 176).

I fail to see why the non-specificity of religion should aid in survival — except perhaps in the survival of religion itself. Secondly, abstract statements about gods are usually bound up with a quite specific religious content. Religions, in fact, do not appear to survive a great deal of social change without themselves going through major change. To place religion at the summit of a social control hierarchy seems nothing short of absurd for me. I fail to see how the meaninglessness of statements about gods can be linked to their adaptive advantage.

> The very qualities of such propositions that lead positivists to take them to be without sense, or even non-sensical, are those that make them adaptively valid (*ibid.*: 176).

Adaptive for whom? Propositions about God have been implicated in some of history's greatest atrocities and in a great share of its day-to-day stupidities. Is it possible that the senselessness of religious discourse, so often described as an opium mystifying the real nature of systems, can be assumed to have any adaptive value at all? Are we really so incompetent that we need religion to ensure our survival? To define religion as adaptive reveals a clear and somewhat disconcerting ideological bias. The foundation of this ideology and its structure can best be understood in terms of Rappaport's and Bateson's attempt to understand why systems do not work.

If systems do not work it is because they are maladapted, i.e. their regulatory hierarchy is somehow mixed up. One prime mover in maladaptation appears to be 'usurpation', where lower order goals take over higher order regulators. In our own system's case this is a question of 'industrialism' which is characterised as production for production's sake.

> High energy technology comes, finally, to serve itself, its goal merely to maintain an ever-increasing flux of energy and material through the techno-social system which it dominates (Rappaport 1977a: 183).

For Bateson, it is a combination of 'technological progress, population increase and certain errors in the thinking and attitudes of occidental culture' (Bateson 1973: 466) that are the 'root causes' (*ibid.*) of the current threat to Man's survival.

It is significant that nowhere is the nature of the social and economic system taken up. As it is the abstract idea level that dominates our system, the maladaptation is linked to the idea that increasing production indefinitely is a desired goal. The solution, then, is quite simple. We must rid ourselves of our 'hubris' — we need a new value system. The accumulation of capital and the need to maintain profit levels in our system is not, however, a question of 'hubris' — the reverse is much more likely.

The major opposition in evolution as Rappaport sees it, occurs when primitive religion is replaced as the dominant structure of social systems, when political and economic goal structures take its place. This opposition corresponds to one between the functioning primitive society in harmony with nature and the expansive, destructive, systems that have developed with the great civilisations and culminated with Western capitalism. This opposition is strikingly similar to that depicted by Rousseau between the Noble Savage, Natural Man, and the industrial-commercial society of his day.

I have attempted elsewhere and in this paper to show that social systems have never been adaptive and that their internal dynamics have always tended to be of an accumulative nature and, finally, that it is this that accounts for the long term development of the destructive and self-destructive system we have today. Not that accumulation in itself is a mistake, but that the social forms it has taken have always tended to be in contradiction with other subsystems upon which its existence depended — this, because it has been a blind process. The religious nature of social systems, the mystified form of all social life, has undoubtedly been a key factor in this disastrous development, not a source of its correction.

What, then, is this ideology all about? It appears as though in the beginning Man was at one with Nature. His activity was regulated in accordance with the cybernetic loops of the larger ecosystem that he was somehow able to grasp through his religion — a religion that must then be envisaged as a lower level manifestation of the larger totality of Nature. Then, with the development of states, Man began to develop and 'control' nature and use if for his own social or, better (in their terms) 'linear' projects, thereby contradicting Nature. This is the problem today and the cause of our present crisis. The elements of this ideology are as follows:

(a) Nature and ecosystems are good — well regulated sets of interlocking programmes, messages and energy flows, all organised in a living system whose highest level is a general purpose programme whose only goal is survival and which, as such, organises all of the other goals and 'projects' in the system that appear as so many subsystems of the larger general purpose programme.

Thus, natural evolution is not a process of selection in which the great majority of species have come to disastrous ends because their programmes were not adjusted to their environments, but the work of a larger Mind — or perhaps ecosystem — a well ordered process of successive adjustments and transformations in which species populations are only 'cogs in the machine'.

It is this interconnected and changing totality that Bateson refers to as MIND — an apparently organised cybernetic system whose successive transformation is described as a learning process. The problem here is how to incorporate everything that is so obviously contradictory and violent in Natural history into this harmonious model — i.e. that species transformation includes the extinction of populations. Bateson's answer lies in his concept of learning which is no more than a 'change of state' — i.e. when I throw a glass against the wall, its shattering is a learning process, also an adaptation to a new environment, self-transformation or self-organisation in the presence of changing conditions. In this way Mind is equivalent to everything that happens in Nature. The rational is real and the real rational in such an approach. Everything that happens appears to be adaptive in such a system. The problem then is how to identify cases of maladaptation.

(b) Mind can be ill as well as healthy. Ill cases are defined as run-away subsystems as opposed to self-maintaining systems. The prime example is our own social system, and it is usually opposed to all of Nature and 'primitive society' which is apparently more Nature than Culture. This attitude is, I insist, a blatant case of Rousseauism, ecologism or what you will. While it is admitted that some systems can be maladaptive, it is only our own industrial ideology and some of its forerunners perhaps that are worthy of accusation.

Ecological anthropology has been at great pains to try to show that what appear to be contradiction-ridden and destructive social systems are in fact adaptive. It is only in this way that it has been possible to explain short stature as an adaptation to lack of protein (Buchbinder: personal communication), to explain warfare as a means of population adjustment, to explain sacred cows as a mechanism of maximisation of protein and calorie production in a half-starving population. The 'primitive' can do no wrong, only we civilised 'conscious purpose' freaks. But then again, more hard-headed and less humanistic ecologists might explain large scale population destruction as a similarly adaptive, i.e. transformative phenomenon. Warfare may alleviate overpopulated regions, economic crises or collapse may reduce the dangers of overheated economies to say nothing of pollution, etc. The approach is obviously vague enough to allow one to arbitrarily decide which system is adaptive and which is not.

I would like to suggest that the ideology of Mind is very much a product of the current crisis of the western sector of the world economy.
The ideological matrix of the ecology movement is very much a part
of this system's basic structure and goes back at least to Rousseau's
anti-civilisation philosophy of the eighteenth century, a philosophy
containing elements of a naturalistic pantheism in which Man is an
organic part of the larger natural world. Rousseau and Hegel have
this one aspect in common. But while Hegel is also a philosopher of
the Totality, he believes in evolution and takes his Mind seriously —
i.e. that there are no mistakes in the real world — the real is always
rational.

The contradiction in the ecological approach is that while they
have a general model of a control hierarchy which they assume to be
the most natural state of affairs, they restrict that model to those
societies considered to be ecologically good. While their cybernetic
model is capable of absorbing all sorts of catastrophies as if they
were smooth transitions, they, for well-founded but ideological
reasons, assume that our civilisation, with its linear thinking and
conscious purpose, is contradictory to the self-maintaining circular
nature of the ecosystem.

While the cybernetic approach is itself a development in mathe-
matics and engineering, the cybernetics of Mind is very much a
result of the crisis of the West as centre of the world system whose
extra-peripheral populations have always provided the model for the
Noble Savage, even if these societies are largely the result of
devolution and even depopulation. Its ideological content has
included such phenomena as 'counter-culture', the use of drugs to
expand 'consciousness', to break down the boundaries between 'self'
and 'world', the belief in the original unity of man and nature and
its disturbance in advanced industrial society. The whole framework
is part of the ideological structure of our system. Its complementary
opposite, which dominates cultural materialism as well as historical
materialism, is the doctrine of progressive evolution which places the
same primitive societies at the bottom of a long and positive development based on that technological growth which is so despised by
the enemy.

It is because the 'ecology of Mind' expresses the underlying
ideology of the system that its solutions are of an ideological nature.
The problem is that we have 'bad' values. We need a new religion,
one that will rid us of the dominance of conscious purpose and bring
us into direct contact with the larger ecosystem of which we are a
part. This political programme is consistent with the world view based

on control hierarchies dominated by rules, principles and gods, where there is no conceivable place for socio-economic structures as such.

Now the idea that industrial growth is good, that the 'business of America is business' is not a 'usurpation' by a lower order of a higher order. The lower order, the capitalist organisation of the economy, is itself the dominant structure of the system. The high valuation of production is its ideological expression and not a regulator of the system. There is absolutely no contradiction in a head of industry indulging in mind-expanding experiences, advanced psychotherapy, etc. In fact, it is very much a privilege of the upper class to partake in such 'revolutionary' activities.

A characterisation of cybernetic anthropology depends on an understanding of the relations between the control hierarchy model, the ideology of Mind and the Rousseauism inherent in the ecological ideology. It appears that for Bateson and Rappaport, the hierarchical ecosystem is always there, but that our own society and all societies based on growth, tend to mess things up – not that the structure itself is altered, but only that it is misused.

The notion of Mind underlies, as an implicit assumption, the major arguments in the cybernetic model. Mind, as we said, is a totality organised as a control hierarchy leading from the lowest forms of life to the systems of ecosystems that make up the living universe and its inorganic energy base – very large indeed and all regulated. We are somewhere between Hegel and Rousseau here, but Hegel is clearly the ideological father of Mind. Hegel's doctrine of the Absolute Spirit is an attempt to eliminate the concept of externality – to reduce all phenomena to parts in a larger system, finite aspects of the higher unity – the evolving Absolute Spirit. The Spirit is equated with God, and individuals are not external but internal aspects of his very being. The conclusions of Bateson are not far off.

> The individual mind is immanent, but not only in the body. It is immanent also in the pathways and messages outside the body, and there is a larger mind of which the individual mind is only a subsystem. This larger mind is comparable to God and is perhaps what some people mean by 'God', but it is still immanent in the total interconnected social system and planetary ecology (Bateson 1973: 436).

This is very much along Hegelian lines – the absorption of individuals into a larger totality of which they are but aspects, where all material as well as mental activity can be defined as 'learning', i.e. activities of the larger Mind. The unity of the totality is a veritable marvel, organised from top to bottom by regular cybernetic processes. The individual's place in all this is quite clear.

As Hegel put it, 'Consciousness of God is God's self-consciousness' (Hegel in Colletti 1973).

The whole philosophy of Mind is based on the *a priori* assumption of harmony except where it is broken by Man's 'conscious purpose'. Mind is a dogmatic construct insofar as harmony is posited as the basis of the entire system and not as an object of investigation. This leads to an overtly religious solution for Mankind in crisis that I find dangerous, to say the least. The innocence is quite astonishing.

> A certain humility becomes appropriate, tempered by the dignity, a joy of being part of something much bigger. A part — if you will — of God (Bateson 1973: 436).

The critique offered here is in no way meant to discredit the numerous brilliant contributions made by Bateson and Rappaport to our understanding of evolutionary processes and mechanisms, social and individual systems, etc. I have concentrated only on one major underlying aspect, the assumption of Mind, which I feel does not alter the contributions made but rather interprets them in a purely ideological way. It is the latter that I find unnecessary in order to understand the relation of primitive societies to their environments, the role of somatic change in evolution, the nature of differentiation in ecosystemic evolution, etc.

Mind is, rather, an overarching framework within which more interesting analyses can be placed — a framework that views systemic processes as ordered by higher instances, the last instance being the 'hand of God'. Evolution thus becomes a well managed programme instead of the contradictory and violent, even though systemic, process that appears on the surface.

Postscript: An Alternative Framework

The principal purpose of the preceding discussion has been to demonstrate the fallacy of the *a priori* assumption of cybernetic regulation in the analysis of social reproduction. This is in no way meant to deny the existence of cybernetic structures but only to locate them correctly in reality. The very general hypothesis suggested here is that while human organisms are indeed cybernetic as individuals, social systems are not. The direction of global activity and of social transformation is a teleonomic result of the interlocking cycles of lower order goal-oriented processes that have little to do with the question of the survival of the larger unit.

There are clearly processes in social systems that have a cybernetic character, whose variables are linked in negative feedback loops. The interest of such *de facto* cybernetic processes depends upon their teleological status. I would suggest that there is a crucial difference between negative feedback processes defined behaviouristically (Rosenblueth, Wiener *et al.* 1943; 1950) and processes that might be designated as purposive (as in Churchman and Ackoff 1950). The former is a descriptive notion that refers to any negative feedback loop in a system. The notion of purposive specifies further that there exists a programme component organising the negative feedback process itself. This means that there exists a sentence in the programme specifying the goal to be attained. It is this property that I take to be crucial in distinguishing teleology from teleonomy. There is a difference between saying that x is the unplanned result of the operation of programme P and saying that x is specified in the programme as the goal to be attained. A thermostat is a servo-mechanism that does not contain a complete programme in itself since it does not specify the temperature goal of the system. This can only be achieved by someone who sets the thermostat, i.e. a programmer. A ritual pig feast is a purposive cybernetic mechanism if there is a programme (existing as a body of discourse) specifying its regulatory function and conditions of action (triggering). Where such conditions do not exist, as I hope to have shown, there may well be a cybernetic process in the descriptive sense, but no purposive cybernetic *activity*. This is true for all such systems whose cybernetic nature is derivable from the structural properties of the system but where those properties are logically independent of the cybernetic result. For example, a business cycle has a cybernetic form, but this form is not programmed as such. Rather, the form is an effect generated by other, logically prior, mechanisms and their interaction over time. To refer to such a phenomena as cybernetic is to use the word as a mere 'cover symbol', a marker that designates a process without providing any insight as to its real nature. It is just this difference between teleology and teleonomy, between programmed and non-programmed cybernetic processes, that is confused by the ecologists of Mind.

The fact that processes can be characterised as cybernetic is due simply to the occurrence of relatively self-maintaining forms, without there necessarily being a programme of self-maintenance. This, however, leaves us with the real problem, to account for such apparently self-regulatory processes without recourse to the cybernetic model which, in such cases, amounts to little more than reified description. The phenomenon to which we are referring is one that has most profitably been explored in more general terms as 'structural stability' (Thom 1972; Prigogine n.d.). Structural stability refers to the maintenance of form in conditions of variable perturbation. It is, as such, a property of structures themselves, one whose limits can be clearly defined. Cybernetic processes are a subset of stable structures. The major difference in general outlook here is that structural stability is not the necessary result of a programmed stabiliser. Instability and stability are in this framework theoretical equals. One is not more 'natural' than the other. A triangular block balanced on one of its apexes is stable within a very narrow range of perturbations (unstable stationary state) while the block, when resting on a side, is truly stable (Prigogine n.d.: 44). Since, however, all stability is subject to fluctuations, the concept is only meaningful in terms of the notion of limits.

Very much of the current ideology concerning the natural necessity of negative feedback is related to the idea that living systems are basically anomalous in nature, disobeying the second law of thermodynamics.

> The maintenance of life would appear in this view to correspond to an ongoing struggle of an army of Maxwell demons against the laws of physics to maintain the highly improbable conditions which permit its existence (Prigogine n.d.: 20).

The work of Prigogine and collaborators (Glansdorff and Prigogine 1971) has today dispelled much of the argument behind this idea. Prigogine has demonstrated that the maintenance of states far from thermodynamic equilibrium is a general pheonomenon not limited to living systems. He has developed an approach in which stable structures emerge from fluctuations far from equilibrium so that evolution is conceived as a phenomenon characterised by the continuous movement from order to instability to new order. Such non-equilibrium structures are called 'dissipative structures', open systems maintained at states far from equilibrium, 'giant fluctuations stabilised by exchange of matter and energy' (Prigogine *et al.* n.d.: 46).

The alternative implied here makes use of a different array of concepts. Instead of regulation, maintenance, negative feedback, Mind we deal with cycles, structural stability, secular trends, constraints, morphogenesis, limits, crisis, 'catastrophe', transformation, etc. In this alternative approach, stability is not the miraculous outcome of self-control but the result of the existence of degrees of freedom within larger systems, of constraints and resistance to perturbation. Stability is a state of affairs and not an organisational principle except when programmed or planned as such. Even in natural science it is becoming increasingly evident that nature is characterised by fluctuations and catastrophic transformations so that there is no reason to limit such phenomena to 'industrial society'. It can be shown that as dissipative structures move further from equilibrium, they reach limits beyond which discontinuous structural change, bifurcation or catastrophe, occurs. The formal properties of catastrophies have been systematically treated by R. Thom (1972) and his results are parallel in many ways to those of Prigogine. The Universe was apparently born in an explosion. When its expansion reaches its limits, the contraction is likely to be just as catastrophic, judging by the current explanations of the almost unintelligible contraction of dying stars into infinitessimal yet infinitely dense 'black holes'. There is, of course, plenty of stability, but this does not warrant the harmonious view of a self-regulating universe.

I would suggest that the cybernetic framework does not adequately take account of the contradictory nature of social or, for that matter, natural phenomena. On the other hand, the framework suggested here, however briefly, and in other works (Friedman 1975; Friedman and Rowlands 1977) does adequately account for processes that are cybernetic in form. It is characteristic of human social systems that they contain many such repetitive cycles. It is conceivable that some social systems might also oscillate between the same values or limits. Social evolution has not, however, exhibited such long run stability. As social systems tend to be of an accumulative nature, stable 'cybernetic' cycles are contained within long term secular trends leading to crises, breakdowns and reorganisation.

References

Ashby, W.R. 1956. *Introduction to cybernetics.* London: Chapman Hall.
——— 1960. *Design for a brain.* London: Chapman Hall.
Bateson, G. 1972. *Steps to an ecology of mind.* New York: Ballantine.
Bertalanffy, L. Von 1968. *General system theory.* New York: Braziller.
Brookfield, H.C. 1973a. Explaining or understanding: the study of adaptation and change. In *The Pacific in transition,* H. Brookfield (ed.). London: Methuen.
——— 1973b. Full circle in Chimbu: a study of trends and cycles. In *The Pacific in transition,* H. Brookfield (ed.). London: Methuen.
Churchman, C. and R. Ackoff. 1950. Purposive behavior and cybernetics. *Social Forces* **29**, 1.
Colletti, L. 1976. *Marxism and Hegel.* London: New Left Books.
Ekholm, K. 1977 (1975). Varför fungera inte samhällen. *Marxistisk Antropologi* **3**, 1.
——— 1976. On the structure and dynamics of global systems. *Antropoligiska Studier* **20**.

Friedman, J. 1972. *System, structure and contradiction in the evolution of 'Asiatic' social formations.* Copenhagen (forthcoming).

—— 1974. Marxism, structuralism and vulgar materialism. *Man (N.S.)* 9, 444–69.

—— 1975a. Religion as economy and economy as religion. *Ethnos* 1–4 (Festschrift for K.G. Izikowitz).

—— 1975b. Tribes, states and transformations. In *Marxist analyses and social anthropology*, M. Bloch (ed.). London: Malaby Press.

—— 1976. Marxist theory and systems of total reproduction. *Critique of Anthropology* 7.

Friedman, J. and M. Rowlands 1977. Notes toward an epigenetic model of the evolution of 'civilization'. In *The evolution of social systems,* Friedman and Rowlands (eds.). London: Duckworth.

Glansdorff, P. and I. Prigogine 1971. *Structure, stability and fluctuations.* New York: Wiley.

Godelier, M. 1973. *Horizon, trajets marxistes en anthropologie.* Paris: Maspero.

—— 1975. Modes of production, kinship and demographic structures. In *Marxist analyses and social anthropology*, M. Bloch (ed.). London: Malaby Press.

Hegel, G.W.F. 1967. *The phenomenology of mind.* New York: Humanities Press.

—— 1972. *The logic of Hegel.* London: Oxford.

Hide, R. 1974. *On the dynamics of some New Guinea Highland pig cycles.* m.s.

Pattee, H. 1973. *Hierarchy theory.* New York: Braziller.

Prigogine, I., P. Allen and R. Herman n.d. *The evolution of complexity and the laws of nature.* m.s.

Rappaport, R. 1968. Pigs for the ancestors. New Haven: Yale.

—— 1971. The sacred in evolution. *Annual Review of Ecology and Systematics* 2.

—— 1971. Ritual, sanctity and cybernetics. *Am. Anthrop.* 73 (1), 59–76.

—— 1977a. Ecology, adaptation and the ills of functionalism. *Michigan Discussions in Anthrop.* 2, 138–90.

—— 1977b. Maladaptation in social systems. In *The evolution of social systems,* J. Friedman and M. Rowlands (eds.). London: Duckworth.

Rosenblueth, A., N. Wiener and J. Bigelow 1943. Behavior, purpose and teleology. *Philosophy of Science* 10.

Rosenblueth, A. and N. Wiener 1950. Purposeful and non-purposeful behavior. *Philosophy of Science* 17.

Thom, R. 1972. *Stabilité structurelle et morphogenèse.* Reading: Benjamin.

Waddington, C.H. (ed.) 1968. *Towards a theoretical biology, Vol. 1.* Edinburgh: Edinburgh University Press.

THE SOCIAL AND ECOLOGICAL RELATIONS OF CULTURE-BEARING ORGANISMS: AN ESSAY IN EVOLUTIONARY DYNAMICS

TIM INGOLD

The culture-bearing organisms of my title are, of course, human beings; and my problem is no less than to reconcile the continuity of the evolutionary process with the qualitative disjunction that separates our supposedly superior selves from the rest of the animal kingdom. My argument is couched within a pair of oppositions: between the social and the ecological, and between the cultural and the organic. These oppositions are not to be regarded as identical, nor even homologous, for the former exists on the level of the encompassing system, whereas the latter exists on the level of the individuals so encompassed. In other words, human beings participate simultaneously in systems of ecological and social relations of production, each as a bearer of a particular constellation of genetically and culturally transmitted traits. The genetic constitution defines the structure of the human being as an organism; the cultural constitution — for the purpose, of the present essay — is defined as a repertoire of technological and ideological models, more or less consciously held, and subject to invention, modification and diffusion quite independently of the biological facts of random mutation and differential reproduction.

Very briefly, my argument will be as follows: that the conjunction of social and ecological systems, at any particular point in time, defines a set of conditions to which the participant organisms are constrained to adapt, either genetically or culturally, or by a combination of hereditary and learned attributes. If the adaptation is largely or wholly genetically programmed, an intensification of economic production beyond the limits of functioning of the ecosystem will necessarily cause the demise of the adapting population, thereby permitting a temporary restoration of equilibrium. If, on the other hand, the adaptation is largely cultural, the population may be able to survive an ecological transformation and hence enter a new phase in a process of social, as distinct from biological, evolution. Culture is therefore a condition, but not the cause, of social evolution, and is in turn conditioned by it.

The Social Relations of Animals and Men

It is fitting to begin with the words of T.H. Huxley, who was the first to demonstrate incontrovertibly the biological continuity between men and apes, and between these anthropoids and the rest of the animal kingdom with which they ultimately share a common evolutionary ancestry. His conclusion was that

> no absolute structural line of demarcation, wider than that between the animals which immediately succeed us in the scale, can be drawn between the animal world and ourselves (1894 [1863] : 152).

Reacting against the prevalent anthropocentrism of his day, Huxley declared that even the so-called 'higher faculties' of sentiment and intellect, traditionally reserved for Man – and only 'civilised' Man at that, 'begin to germinate in lower forms of life' (*idem.*). If we are prepared to admit that the individual human being grows from an egg indistinguishable from that of the dog, so Huxley argued, we must surely realise the futility of attempting to draw a rigid distinction, on either the anatomical or the psychical plane, between the human species and the rest of the animal kingdom.

Few today would disagree with the argument that Man has evolved, in conformity with the Darwinian theory of variation under natural selection. Yet anthropology has remained stubbornly insistent in its belief in a quantum-jump from Nature to Culture, variously attributed to the manufacture of tools, the development of language, and the institution of an incest taboo. As Kroeber wrote in his celebrated paper on 'The Superorganic':

> The dawn of the social . . . is not a link in any chain, not a step in a path, but a leap to another plane. It may be likened to the first occurrence of life in a hitherto lifeless universe . . . (1917: 209).

Thus, inorganic is to organic as organic is to superorganic. Now by 'social', Kroeber meant in fact those technical and intellectual capabilities that we would call 'cultural': indeed, he appeared to treat the social and the cultural more or less synonymously. It is essential, for the thesis I wish to present, to dispel this confusion. Society is not the sum of achievements of past generations preserved in the present, the cultural equivalent of the gene pool. It is, on the contrary, a system of relations, with a reality quite independent of its particular intellectual representations. We have therefore to consider two possible kinds of discontinuity: between ecosystem and social system, and between the genetic and cultural transmission of individual attributes.

Having made this distinction, we can proceed to posit that the domain we call 'social' exists, and has existed, prior to the development of culture. Men did not, for example, become social because they developed the capacity of speech; but rather the capacity of speech developed because men were social (Engels 1934: 173). The same argument applies to the construction of tools and the elaboration of sexual prohibitions. Culture could not have emerged, fully formed, from nowhere; but must have developed under the influence of a set of conditions that were social as well as ecological in character. A corollary of this argument is that both kinds of condition must also act in conjunction upon genotypic variants in a population. I am suggesting, then, that Huxley was correct to view the evolution of the human intellectual faculty as a process admitting no absolute discontinuity; but that this entire process has been governed by the interplay of two quite distinct kinds of system, social and ecological, whose divergence lies in the origins of the animal kingdom itself. As I shall show later, 'selective pressure' can be defined only in terms of the articulation between these two systems and is an indication of the tendency of the former to 'drive' the latter towards the limits of its functioning.

This suggestion implies a double criticism: of cultural anthropology for mistaking a disjunction between biological and social domains of both animal and human existence for a temporal leap from one domain to the other, supposedly separating Man from the animals; of Darwinian theory for imputing to the social order of animals a kind of natural capitalism. As Engels wrote, in a letter to Lavrov:

> The whole Darwinian teaching of the struggle for existence is simply a transference from society to living nature of Hobbes's doctrine of 'bellum omnium contra omnes' and of the bourgeois-economic doctrine of competition together with Malthus's theory of population (cited in Schmidt 1971: 47, cf. Sahlins 1976: 101 ff.).

Yet it will be objected, at once, that I have reified the concept of Society. What does it mean to say that animals are social, when their behaviour is largely patterned by instinct? My answer is that both animals and men are engaged in a process of economic production, in the sense that they act wilfully on their environments in order to obtain the means to support and reproduce themselves. Therefore, they are involved in social relations of production, irrespective of whether the force of labour upon its subject matter is genetically or culturally mediated.

This point may be demonstrated most clearly if we digress, briefly,

on the phenomenon of taming. It is a fact that men have succeeded
in establishing their mastery over individuals of diverse animal
species, which are compelled to contribute their labour towards the
satisfaction of human wants. Are they not, then, involved with men
in social relations of domination, akin to those of slavery? The
question was taken up by Marx, who answered decisively in the
negative, on the grounds that animals lack will:

> Appropriation [of another's will] can create no such relation to animals
> . . . even though the animal serves its master . . . Beings without will,
> like animals, may indeed render services, but their owner is not thereby
> lord and master (1964: 102).

Similarly, in *Capital*, domestic animals are classified alongside
primitive tools as instruments of labour (1930: 172). But this is
to relegate animals to the status of mindless, power-driven machines.
In truth, the domestic animal is no more the physical conductor of
its master's activity than is the slave: both constitute labour itself
rather than its instruments and are therefore bound by social relations
of production. In other words, taming is not a technological
phenomenon.

But are not domestic animals at least in part human creations,
fashioned by men to their requirements, as they might fashion tools?
And are they not therefore the products of labour rather than agents
of production? To these objections I would reply firstly, that neither
the taming of animals, nor the attachment to them of man-made
instruments, has anything to do with their modification through
selective breeding, although all are commonly confused under the
concept of 'domestication'. Secondly, even when men breed their
tame animals, they merely act upon successive variations already
brought forth in the course of natural reproduction: in no wise do
they create these variations (Darwin 1859: 26). And finally, if the
upkeep of domestic animals requires an investment of human
labour, so too does the upkeep of children, yet this does not turn
children into tools. From my conclusion, that the relation of
taming is essentially *social*, it is but a short step to the inference
that animals are likewise involved in social relations with their own
kind. This at once invalidates the criterion by which Marx separates
Man from the animal kingdom.

Ecological and Economic Production

> Men . . . begin to distinguish themselves from animals as soon as they begin
> to *produce* their means of subsistence, a step which is determined by their
> physical constitution. In producing their means of subsistence men

indirectly produce their actual material life (Marx, in Bottomore and Rubel 1963: 69).

Evidently, everything hinges on the concept of production. I find it necessary to distinguish two senses of the term, one ecological, the other economic. Ecological production refers to the thermodynamic process whereby energy from the sun fuels the creation of organic material in Nature. Economic production, on the other hand, refers to the expenditure of labour, whether by animals or men, in order to obtain from Nature the means of subsistence.

Now it need hardly be said that ecological production is going on in all living things, for it is none other than the life-process itself. Imagine, for example, a part of an ecological system — a particular food chain — linking grass, a herd of herbivorous game, and a group of human hunters. The grass assimilates solar energy by photosynthesis, and this is converted, with a certain efficiency, into potential energy stored in plant tissues: quite simply, the grass grows. The herbivores graze the grass, assimilating a proportion of that potential energy, which is converted — again with a certain efficiency — into energy stored in the flesh of the living animals. The men, in turn, kill the animals and consume the meat, facilitating a further step in ecological production: the growth and multiplication of human bodies.

Clearly, given the requisite nutrients, and a source of energy in sunlight, grass need do nothing itself in order to grow. But the same is not true of either the herbivores or the hunters. Both must perform a certain amount of work in order to secure a flow of energy and materials from their food resource into their own bodies. This work includes movement, from one food location to the next, the extraction of the food, and its consumption, preceded perhaps by intermediary stages of preparation. In the case of the herbivore, all these acts may appear to be proceeding simultaneously: as the animal uses its legs to move over the pastures, it cuts the blades of grass with one set of teeth, and chews with another. Extraction and consumption are immediate in space and time, whereas among human hunters they are at least to some extent separate (Lee 1969: 49). But this does not mean that the animal economy can be reduced to ecological relations of production. It is a common error to confuse the consumption of food with the growth of living bodies, when these are in fact two entirely separate processes: one taking place through wilful actions by the subject, the other by organic reactions within the subject. Or, to restate the point in more obvious terms: animals must eat in order to grow, but eating and growing are not the same.

What I have called ecological and economic production are there-fore autonomous but complementary. Wherever there are resources to harvest as food, ecological production must be going on; yet ecological production in the consumer population depends upon action by those consumers to obtain their food and bring it to the point where it may be organically assimilated. Now, it is most important to recognise that the energy expended in the successive stages of economic production (movement, extraction, preparation, consumption) bears no relation whatsoever to the calorific content of the food. When the hunter throws a spear to kill his game, muscular energy is converted to kinetic energy in the motion of the spear, all of which is ultimately expended in the friction of impact. There is no more relation between the energy needed to throw the spear and the energy content of the game than there is between the electric current required to trigger a thermostat and the heat generated by the boiler it controls. Pressing the analogy, we might regard the action of the hunter in throwing his spear as functioning to 'switch off' a process of ecological production in the game, and his action in processing and eating the meat as functioning to 'switch on' a process of ecological production in his own body.

It follows that it is meaningless to define the efficiency of economic production in thermodynamic terms, as a ratio of energy expended in obtaining food to the calorific content of the food produced. Of course, these quantities could be measured in theory (though guessed in practice), and a ratio could be calculated, as Harris has attempted to do in devising his index of 'techno-environmental efficiency' (1971: 203ff). But the resulting figure does not characterise any real, physical process of energy conversion. It is only in a social sense that we can speak of labour being em-bodied in products. Hence, the values assigned to labour and to the products of labour must be socially rather than physically defined. Productive efficiency can then be estimated as the ratio of these values:

> The [economic] productivity of a system will be the measure of the ratio between the *social* product and the *social* cost that it implies. In so far as production operations combine quantifiable realities (resources, instruments of labour, men) and require a certain time to be completed, qualitative, conceptual analysis of a system leads on to numerical calculation (Godelier 1972: 265, my emphases).

Thus, the distinction between economic and ecological production is a logical corollary of that between the social system and the ecosystem, and appears not with the emergence of Man from his prehominid ancestors, but with the emergence of social, tameable species of animals from lower forms of life.

Now in my terms, the gathering or harvesting of food from Nature, whether by collectors or cultivators, represents an aspect of economic production. Yet it is customary to contrast food-collection with food-production, attributing the latter to Man alone, or more particularly to human cultivators and pastoralists. Thus for Engels:

> The most that the animal can achieve is to *collect*; man *produces*, he prepares the means of life, in the widest sense of the words, which without him nature would not have produced. This makes impossible any un-qualified transference of the laws of life in animal societies to human society (1934: 308).

Later on, the archaeologist V.G. Childe was to coin the same distinc-tion to describe the transition from palaeolithic 'savagery' to neolithic 'barbarism', which he regarded as 'an economic and scientific revolution that made the participants active partners with nature instead of parasites on nature' (1942: 55).

Though identical in idiom, these respective formulations are fundamentally discrepant in content. For whereas Childe equated the inception of production with the domestication of plants and animals, Engels placed it much further back in human evolutionary history, at the point where hunters and collectors began to construct their own implements.

Man the Toolmaker

> The specialization of the hand – this implies the *tool*, and the tool implies specific human activity, the transforming reaction of man on nature, production (1934: 34).

It is unnecessary here to go into the question of whether or not Man is, in fact, unique in his ability to construct tools by cultural design. Engels's proposition is open to the more general objection that relations of production are not themselves generated by technological advance (Friedman 1974: 450). If rudimentary tool-making occurs together with a certain form of society, it does not follow that the one caused the other. On the contrary, the system of social relations must already exist for the corresponding technological development to take place at all. Just as the steam-engine was a product of capitalism, or large scale irrigation works a product of the archaic state, so surely, the stone-chopper was a product of 'hunting and gathering' relations of production. In each case, of course, technological innovation may have accelerated the development of the corresponding social system, but it did not bring that system into existence. In short, technology, is not the 'prime mover' behind human social evolution, nor is it subject to some pre-ordained order of progression (cf. Harris 1968: 232).

I would argue, rather, that the design and manufacture of tools is guided within an underlying set of ecological and social conditions, which determine the adaptedness or otherwise of alternative models. But tools, defined as cultural artefacts, hardly comprise the main part of the hunter's or gatherer's equipment. Far more important is his knowledge about the different species of animals and plants available to him, where they may be located, how they reproduce, and how — in the case of animals — they may best be approached and captured. And again, for certain kinds of hunting which involve the co-operation of a number of people, an effective allocation of tasks among the participants may be more crucial for the success of the hunt than the sophistication of their material equipment. The adaptive repertoire of hunters and gatherers must therefore be broadened to include such immaterial factors as skills, organisational techniques, and knowledge about the natural world and the way it behaves. As Sahlins writes, 'a technology is not comprehended by its physical properties alone' (1972: 79).

Reasoning thus, the boundary between native technology and cosmology becomes indeterminate. Taken together, they comprise a body of conscious models that individuals carry in their heads, and communicate symbolically. If any distinction can be drawn, it is between models *of*, and models *for*, between representations of reality and instructions for action: yet it is characteristic of human symboling that these kinds of models are inter-transposable (Geertz 1966: 7–8). Their adequacy may be assessed according to whether or not they motivate individual producers to perform in a manner that is materially appropriate, given the social and ecological conditions within which they are constrained to operate (cf. Rappaport 1968: 239). I take issue here with those who appear to equate, or at least to confuse, technological and environmental relations under that awkward hybrid, the 'techno-environment'. Technology is a corpus of culturally transmitted knowledge, expressed in manufacture and use, and as such it belongs with ideology in the domain of the super-structures. The 'forces of production' consist not of tools, nor of their connections with men, but of the physical relations that men establish with the natural environment through the mediation of their ideas and techniques. On the infrastructural level of the mode of production, the social is dialectically opposed not to the technological but to the ecological.

But I take issue equally with the cultural materialist argument that regards the social system as a superstructural epiphenomenon of the relation between environment and technology (e.g. Harris 1968). This position rests on a fundamental confusion between the *social* relations of production and the *technical* organisation of work, a confusion shared by some Marxist writers (e.g. Terray 1972). Patterns of co-operation, along with skills and equipment, form a part of the means whereby a population adapts to its environment (Steward 1955: 40–1); but the objectives of this adaptation can only be defined in terms of the inner logic, or 'rationality' of the social system in which that population is involved. The fallacy of cultural materialism, as Friedman (1974) has shown, is to take a correspondence between technological and social systems for a principle of positive determination.

Yet Friedman is just as mistaken to ascribe to technology the power of negative determination. Consider the following proposition:

> A given technology in given environmental conditions constitutes a techno-ecological system whose internal properties impose a certain number of constraints on the functioning of the productive relations by determining the outer limits of technical reproduction of a population at a given level of productivity (1975: 166).

To this I would object that the 'techno-ecological system' is as monstrous a construct as its companion, the 'socio-cultural system'. The functions of technology, as of institutional and ideological aspects of culture, are laid down by the existing social relations of production and ecological conditions of reproduction. In other words, the properties of the technological system are not autonomous but derivative: they may retard or accelerate development on the level of social relations, but they do not impose outer limits on this development. Such constraints are a property of the ecosystem alone.

Let me cite, perhaps a little unfairly, a sentence from a recent paper by an economic anthropologist, for it epitomises so well the kinds of confusion that it is my purpose to dispel:

> At the point where the co-operatively organized human economy separated out from the organism-centric realm of animal energetics, the construction and manufacture of equipment and tools for future use was present (Cook 1973: 46).

I have tried to show that the productive activity of animals can no more be reduced to 'energetics' than can that of humans. To repeat:

the energy expended in the extraction and consumption of food is in no way embodied in its calorific content. The same, of course, is true of energy expended in the manufacture of tools, which have no calorific content. It is not tool-making that introduces an economic dimension into production, but rather the capacity to engage in planned, purposeful action – a capacity that is certainly not unique to Man. Moreover, the construction and use of equipment, and co-operative organisation, which Cook appears to equate with the forces and relations of production respectively, are merely two aspects of a generalised mode of cultural adaptation whose determinants are specified by the intersection of encompassing social and ecological systems.

It may readily be agreed that economic production requires – besides labour itself and the subject matter of labour – a set of instruments. But these instruments may be fashioned according to a template that is either genetic or cultural, and may be either an inseparable part of the body or detachable from it (Marx 1930: 169–70; cf. Engels 1934: 34; Sahlins 1972: 79–80). The beaver fells trees with its teeth, Man fells with an axe, but if in each case the activity is wilful, then both the tooth and the axe are instruments. In the course of evolution, many of the instrumental functions of human teeth have been transferred to culturally-fashioned, detachable tools such as knives and grinding-stones. Yet human instruments are not exclusively cultural: we still use our teeth for eating and our legs for locomotion, even though cultural substitutes are now available. This fact does not in the least detract from the essentially social character of these activities. We may suppose, moreover, that for the greater part of Man's evolutionary history, innate technical capacities played a very much larger part in the assurance of survival than they do today.

That both organic and cultural adaptation have proceeded simultaneously, under social and ecological conditions of the same order, is evidenced by the concurrent development of stone tools and the hands that fashioned them. Engels was clearly aware of this, but his understanding was blurred by an apparent belief in the inheritance of acquired characteristics, which led him to assert that the human hand 'is not only the organ of labour, *it is also the product of labour*' (1934: 172). We now know, of course, that changes in the hereditary constitution of men, as of animals, result not from the conscious design of past generations but from random genetic mutation. It is this that critically differentiates organic from cultural adaptation. Yet it is clear that a close interdependence exists between these two processes, for selection apparently favours those individuals physically better able to make use of available cultural techniques. Thus the 'precision grip' of the human hand evolved in response to the advantages conferred by the use of ever more sophisticated tools. As Napier suggests, 'the stone implements of early man were as good (or as bad) as the hands that made them' (1962: 62). In short, it is incorrect to suppose that the process of cultural adaptation began where that of organic adaptation left off (Geertz 1962: 718–25).

But the assumption that tool-making is necessarily a culturally transmitted attribute is itself open to question. Haldane offers the following provocative comment:

> During the lower palaeolithic period, techniques of flint chipping continued with very little change for periods of over 100,000 years. It seems to me possible that they may have been as instinctive as the making of spider's webs, even if most flint chippers saw other men chipping flints (1956: 9).

The issue here is not so much whether flint chipping was more or less instinctive, for in principle we can never know, but rather that any piece of behaviour will contain genetically and culturally transmitted components, and will be selected upon on the basis of infrastructural social and ecological criteria, the former defining the use to be made of the environment, the latter the limits of viability. In the course of human evolution, cultural transmission has assumed an increasingly dominant role, to the extent that hereditary attributes have become almost immune to direct selective pressures. Nevertheless, it is impossible to define any kind of qualitative jump from the instinctive to the symbolic, for the conscious blueprint (model *of*) is never more than a partial representation of subsequent performance. The 'great evolutionary divide' (Sahlins 1972: 80) is a figment of anthropocentric imagination.

Up to this point, I have been concerned to stress the continuity from animals to men, whilst recognising that this continuity is grounded in a fundamental disjunction between social and ecological systems. Yet Man is surely unique insofar as his productive history has been marked by a series of qualitative transformations in the material conditions of existence. Perhaps the most important of these was the inception of cultivation, or 'food-production' in the sense coined by Childe, a sense which remains in conventional usage today. I wish to argue, now, that although economic production, as I have defined it, is common to both animals and men; the cultural transmission of technique is a necessary condition for the emergence of cultivation, and hence for the process of social evolution that was predicated upon it. Let me return for a moment to Engels, who

recognised as clearly as any modern ecologist the reciprocal nature of
the interaction between organisms and their environment:

From Gathering to Cultivation

> Animals . . . change their environment by their activities in the same way,
> even if not to the same extent, as man does, and these changes . . . in turn
> react upon and change those who made them. In nature nothing takes
> place in isolation (1934: 178).

But if that is so, how does the impact of Man differ from that of
animals? Is it merely a question of degree? Engels continues:

> But animals exert a lasting effect on their environment unintentionally
> and . . . accidentally. The further removed men are from animals, how-
> ever, the more their effect on nature assumes the character of pre-
> meditated, planned action directed towards definite, preconceived ends
> (*idem.*).

This is surely to underestimate the mentality of animals, and to over-
estimate the ability of men to predict the consequences of their
actions. Engels is here confusing the immediate motivation for
behaviour with its long-term effects. Both men and animals act
wilfully, in accordance with some specific rationality, but within
constraints imposed by their natural environments. And with both
men and animals, these actions may have long-term environmental
consequences which are not willed by them.

Later in his argument, Engels himself admits that animals engage
in rational action. In apparent contradiction to the contention
cited above, he suggests

> that it would not occur to us to dispute the ability of animals to act in a
> planned, premeditated fashion. . . In animals the capacity for conscious,
> planned action is proportional to the development of the nervous
> system, and among mammals it attains a fairly high level . . . But all the
> planned action of all animals has never succeeded in impressing the stamp
> of their will upon the earth. That was left for man (1934: 179).

So, after all, the question is one not of the existence of rationality
but of the ability to construct an artificial environment. Man is the
only animal to engage in the deliberate destruction of natural
vegetation in order to replace it with an ecologically more productive
plant cover of his own choosing. This is a kind of environmental
modification quite different from the long-term degradation that may
stem from excessive pressure on resources under any regime of plant
exploitation, whether practised by men or by herbivorous animals.

This, too, must be what Engels meant by the contrast between 'use' and 'mastery':

> In short, the animal merely *uses* its environment, and brings about changes
> in it simply by its presence; man by his changes makes it serve his ends,
> *masters* it. This is the final, essential distinction between man and other
> animals, and once again it is labour that brings about this distinction (1934:
> 179–80).

Yet this 'mastery', if such it is, appeared but late in human evolutionary
history. Far from setting men apart from animals, it merely separates
'neolithic' men and their successors from hunters and gatherers, human
and non-human alike. Moreover, there is good reason to believe that
the inception of cultivation was not contingent upon human technological and intellectual advance, that it was no new 'discovery',
but rather that it was the direct outcome of an intensification of
economic production beyond the environmental capacity under the
gathering mode.

Rudimentary cultivation, after all, requires no more advanced
technology than the axe, digging-stick, and knowledge of fire, all of
which were available to 'palaeolithic' gatherers. For Childe, the
critical innovation of 'food-producing' lay in the science of
selective breeding, but there are no grounds for assuming that
either plant or animal husbandry should be accompanied by artificial
selection (Higgs and Jarman 1972). What, then, is the source of the
discontinuity between gathering and cultivation? The answer lies
not in the breeding of plant domesticates but in the element of
ground-preparation. By burning woody vegetation, turning the soil,
or flooding it with mineral-bearing waters, cultivators increase the
rate at which nutrients are restored to the topsoil, and hence also
the rate at which plant food may be extracted per unit of land area
(Geertz 1963: 12–37). This, in turn, allows a higher density of
human population to be supported but only at the cost of an
additional labour input expended in such activities as plot-clearance,
planting and weeding. The more dense the population working the
land, the harder they must work to feed themselves (Boserup 1965).
If they are to produce a surplus to support a non-producing elite,
they must work harder still. It follows that a major contributory
cause of the transformation from gathering to cultivation, at least
of staple food crops, must lie in the pressure of population on
limited environmental resources, and that people would not
necessarily cultivate unless subject to such pressure, even if they
were familiar with the requisite technique (Binford 1968; Cohen
1977).

Now, ground-preparation requires not only an input of labour in expectation of a delayed return, but also the separation in space or time of the complementary activities of planting and harvesting. Whereas for the gatherer a crop unharvested is equivalent to a crop planted, the cultivator must reserve a portion of the harvest for replanting (Zohary 1969). Consequently, the inception of cultivation entails new social relations of production, which establish control by solidary groups over the fields they have laboured to prepare and control within each group over the storage and distribution of the crop (Meillassoux 1972). It is these social relations, rather than new techniques, which provide the impetus towards population growth and surplus production under cultivation. Population growth is thus both cause, in forcing more intensive use of the land, and effect, in that material prosperity under conditions of sedentary agriculture depends upon the reproductive recruitment of labour. The ramification of systems of cultivation may be viewed more as an accommodation to this population growth, or its 'overflow' to surrounding regions, than as a diffusion of technological innovations (Flannery 1969).

In short, the transformation from gathering to cultivation occurs at the point where the component social and ecological systems of the gathering mode of production 'strain to the limits of functional compatibility' (Friedman 1974: 499), setting up an inter-systemic contradiction which is resolved by the coming into being of a new set of relations, both social and ecological (Godelier 1972: 90). The environment, under such circumstances, plays an ultimately determining role in specifying the limits of the old and the conditions of the new. But there is no reason why this kind of inter-systemic contradiction should afflict only human populations. Consider this very general statement by an evolutionary biologist:

> By accepting rate of increase as the measure of population adaptation, one is forced to conclude that the evolution of almost every group is characterized, sooner or later, by diminishing adaptation to its environment (Stern 1970: 60).

To put it another way, the resolution of one contradiction is bound to lead in due course, and under its own momentum, to the emergence of another. Why, then, have other animals not coped with the same problem by turning to cultivation? To answer this question, we must return to the contrast between the processes of organic and cultural adaptation.

Organic and Social Evolution

How men participate in any ecosystem depends . . . upon the cultural

baggage of those who enter it, what they and their descendants subsequently receive by diffusion or invent themselves (Rappaport 1971:
246).

It is commonplace to invoke parallels between the genetic phenomena
of mutation and drift, and the cultural phenomena of invention and
diffusion, as the above passage implies. Yet the differences are equally
crucial. Genetic mutations are created entirely at random. Once
established in a population, they cannot be eliminated save by a
failure of those organisms bearing them to leave viable offspring, nor
can they spread save by a proliferation of offspring. Selection is thus
absolutely linked to differential reproductive success. Invention or
conjecture, on the other hand, takes place only in response to a perceived problem. It is true that an element of chance may be involved,
but even the initial retention of an idea requires that it should appear
to have some potential utility as a solution to that problem.

Alternative inventions or conjectures then become subject to a
process of trial-and-error, in the course of which some may be
rejected as unsatisfactory whilst others may be widely adopted in
place of previously existing solutions. The same may be said both of
technological models *for*, and of cosmological models *of*, since in
each case available alternatives will be tested for their efficacy in
either explaining or acting upon the real world, in accordance with a
set of premises that are socially given. Now there is, of course, an
analogy between this kind of error-elimination and the operation of
selection on genetic traits, an analogy that has been stretched almost
to the point of absurdity by Popper:

> The growth of our knowledge is the result of a process closely resembling
> what Darwin called 'natural selection'; that is, *the natural selection of
> hypotheses*: our knowledge consists, at every moment, of those hypotheses
> which have shown their (comparative) fitness by surviving so far in their
> struggle for existence; a competitive struggle which eliminates those
> hypotheses which are unfit (1972: 261).

But the difference, which is essential to my argument, is that conjectures and inventions which cease to be adaptive may be eliminated
otherwise than by the elimination of those holding them, or more
significantly, by their failure to reproduce. Our theories and hypotheses, in Popper's words, 'perish in our stead' (*idem.*). Were this not
the case, social evolution would be an impossibility.

A fundamental premise of the Darwinian theory of natural
selection is that although all organisms have the potential to multiply
at a geometrical rate of increase, their actual abundance will be held
within finite, environmentally determined limits (Darwin 1859:

95–6). If this were not so, variants of a population endowed with advantageous traits would merely reproduce alongside those less advantaged, without actually displacing them, generating a cumulative chaos of variation. Now, the regulation of numbers around a balance point implies the operation of some form of density-dependent reaction, such that environmental checks on natural increase become more severe as population density rises and less severe as it falls. The most obvious of these checks, and the one that figured most prominently in Darwin's writings, is a function of direct competition for food. But a Malthusian mechanism of this kind, if not anticipated by some other factor such as predation or intra-specific territoriality, tends to be highly imperfect, coming into effect some time after the optimum density has been exceeded, only to reduce numbers far below this level. Consequently, populations subject to this form of control tend to undergo more or less pronounced oscillations of growth and decline.

Consider an animal population whose numbers are fluctuating on a regularly cyclical pattern. When the rate of growth is at a maximum, the intensity of selection will be at a minimum, so that new variants capable of surviving in a competition-free environment will reproduce without either displacing, or being displaced by, previously existing ones. Conversely, selection will be most intense during the period of maximum decline, when only the better advantaged variants produce enough offspring for replacement, the less advantaged being eliminated altogether. In other words, during each phase of growth, variants are introduced which are subsequently selected upon during the following phase of decline. Viewing a large number of such cycles in series, we may observe a continuing process of adaptation, or what we call organic evolution. But this process is more than the mere achievement of a qualitative 'fit' between the properties of the animal species and those of its environment. Variants are selected not only for their ability to survive in the presence of particular adverse factors, but also on the basis of their relative reproductive success in competition with other variants in the population.

This distinction defines two contrasting modes of selection, which may be termed environmental and competitive respectively (Nicholson 1960). Adaptation under environmental selection tends towards a limit of perfection, at the point when every individual is immune to the adverse factors which impinge upon it. Any further development beyond this limit must depend upon a prior change in the conditions of the environment. Adaptation under competitive selection, to the contrary, tends towards no upper limit, but can proceed indefinitely, for the standard of selection necessarily rises in step with increased reproductive potential. The more adapted the individuals of a population, in terms of comparative fitness, the harsher the competition to which they are subjected. Yet this competition, as I have shown, is a function of the density-dependent control of population numbers. Hence we reach the fundamental conclusion that the progressive adoption of novel traits, coupled with the elimination of those that are relatively inferior, depends upon the continued operation of those forces that act to maintain a dynamic equilibrium in the relations between a population and its environment (Stern 1970: 59).

Let me restate this conclusion in another way. The growth of an animal population leads to the intensification of economic production, which may eventually proceed to the limits of functioning of the ecosystem in which that population is involved. The end of each growth phase is therefore marked by an emergent 'contradiction' between social and ecological systems, which is resolved by an eventual reduction in population that returns both systems to their points of origin. It is this reduction that brings the pressures of competitive selection to bear, eliminating those variants that cannot withstand the crisis, and promoting others whose greater reproductive potency only increases the rate of subsequent growth, without in any way overcoming the limits to growth. Since selection is here conditional upon the homeostatic adjustment of population to resources, it follows that organic adaptation takes place within a framework of ecological and economic production that remains qualitatively in a steady state. There is, therefore, no social evolution.

Consider now an expanding population of humans, exploiting its environment by means of attributes that are, in the main, culturally transmitted. Far from becoming fixed at random, these attributes are developed and perfected by conscious design, so that traits possessing negative utility may be eliminated whilst the growth of population remains relatively unconstrained by environmental factors. One consequence is that the process of cultural adaptation is very much more rapid than that of organic adaptation. But still more important is the fact that, when the environmental limit is approached, the population may modify its cultural practice to accord with transformed social and ecological conditions, and so continue to increase as fast, if not faster, than before. This is possible only because the transmission of cultural means is separable from the biological process of reproduction. Yet on account of this possibility, an analogue of competitive selection, which rests on the premise of density-dependent control of numbers, cannot be applied to the realm of cultural phenomena.

Cultural adaptation is analogous, rather, to organic adaptation under the environmental mode of selection. This conclusion has two major implications that stem from the distinction outlined above between environmental and competitive modes. Firstly, adaptation will tend towards a limit defined by the conditions of survival. Secondly, the intensity of selective pressure will diminish with increasing adaptation and will be at its peak at the point when the population is least well fitted to its environment. Now this point, just as in the case of a non-human population, is marked by the onset of functional incompatibility between pre-existing social and ecological systems. But the resolution of this 'contradiction' takes a fundamentally different course. Each component system, rather than being returned to its point of origin, is qualitatively transformed, yielding a new set of social and ecological parameters of growth. Selective pressure is generated, therefore, not by the intensification of competition for diminishing returns within an unchanged environment, but by the discrepancy between the pre-existing pattern of adaptation and transformed environmental conditions, which throw up new problems of survival. The solution to these problems need not involve the invention of new tools or techniques, for the population may already carry a number of cultural 'pre-adaptations' to the changed conditions of livelihood. But it does require new ways of relating existing technologies to environmental possibilities.

To sum up: whereas organic evolution under competitive selection proceeds indefinitely over a series of cycles of growth and decline, cultural adaptation under environmental selection proceeds towards a limit within each phase of growth. But instead of a steady-state oscillation, we have a series of such growth phases, each characterised by specific social and ecological parameters, and each initiated by a transformation in the infrastructural conditions of production. It is to the succession of phases and transformations on the level of economic production that we refer by the term 'social evolution'. I should stress that organic and social evolution, as they have been defined here, are processes of totally different kinds, operating on different planes. No analogy can be posited between them, akin to that between the selection of genetic and cultural traits. The evolution of society is not governed by a selective process, for what we call 'selective pressure' is itself an indication of the development of the social system beyond the environmental limits. In other words, the premises of selection can only be defined in terms of the conjunction of social system and ecosystem, a conjunction that contains within itself the conditions of its future development. The course of social evolution is therefore predictable, whereas the course of organic evolution is not. One may foresee the problems, but the range of solutions is potentially infinite. The principle of adaptive advantage, as Sahlins has pointed out, 'is indeterminate: stipulating grossly what is impossible but rendering suitable anything that is possible' (1969: 30).

My argument is now complete. In the most general terms, I conclude that where individual attributes are genetically transmitted, a population will be unable to adapt to a radically transformed environment whilst maintaining its own numbers. Impending contradictions between social and ecological systems must therefore be resolved by a return of both to their starting points, yielding in the long term a series of repetitive cycles on the infrastructural level whose effect is to promote, on the superstructural level, a continuing process of organic adaptation under competitive selection. But if individual attributes are culturally transmitted, a population may survive an environmental transformation by making appropriate technical adjustments, such as to bring into existence a new and qualitatively different system of social relations. Each such transformation marks a phase in a determinate process of social evolution, whilst within each phase there will be a trend of cultural adaptation. This trend, like that of organic adaptation, is indeterminate; but the retention of innovations is not random, nor does the elimination of inferior traits depend upon the operation of a homeostatic population regulator. Consequently, the Darwinian image of the 'struggle for existence' is not appropriate to describe the mechanism by which cultural attributes are selected. Cultural adaptation, unlike organic evolution, works towards finite targets.

I hold, with Huxley, that Culture is one with Nature, having developed under the same kinds of conditions that have governed the evolution of organic forms. I reject, therefore, the view of Culture as Superorganism, and subscribe to the premise underlying both the functionalism of Malinowski and contemporary ecological anthropology, that 'Culture, through its relation to culture-bearing organisms, remains ultimately subservient to laws governing living things' (Rappaport 1971: 24). Where I differ is in my insistence that these laws are social as well as ecological. In stressing the continuity between human and animal economies, I am not trying to reduce the human economy to ecological relations of production but am concerned to give due recognition to the economic dimension of production in animal societies. To those who have not already despaired at the manifold inconsistencies in my argument, let me end with an exhortation. Can we not cease to biologise Man, and turn instead to the socialisation of animals?

References

Binford, L.R. 1968. Post-pleistocene adaptations. In *New perspectives in archaeology*, S.R. Binford and L.R. Binford (eds.). Chicago: Aldine.

Boserup, E. 1965. *The conditions of agricultural growth*. London: Allen and Unwin.

Bottomore, T.B. and M. Rubel (eds.) 1963. *Karl Marx: selected writings in sociology and social philosophy*. Harmondsworth: Penguin.

Childe, V.G. 1942. *What happened in history*. Harmondsworth: Penguin.

Cohen, M.N. 1977. *The food crisis in prehistory: over-population and the origins of agriculture*. New Haven: Yale University Press.

Cook, S. 1973. Production, ecology and economic anthropology: notes towards an integrated frame or reference. *Soc. Sci. Inform.* **12** (1), 25–52.

Darwin, C. 1859. *The origin of species* (1950 reprint). London: Watts.

Engels, F. 1934. *Dialectics of nature*. Moscow: Progress.

Flannery, K.V. 1969. Origins and ecological effects of early domestication in Iran and the Near East. In *The domestication and exploitation of plants and animals*, P.J. Ucko and G.W. Dimbleby (eds.). London: Duckworth.

Friedman, J. 1974. Marxism, structuralism and vulgar materialism. *Man (N.S.)* **9**, 444–69.

——— 1975. Tribes, states and transformations. In *Marxist analyses and social anthropology*, M. Bloch (ed.) (A.S.A. Studies **2**). London: Malaby.

Geertz, C. 1962. The growth of culture and the evolution of mind. In *Theories of the mind*, J.M. Scher (ed.). New York: Free Press of Glencoe.

——— 1963. *Agricultural involution: the process of ecological change in Indonesia*. Berkeley: University of California Press.

——— 1966. Religion as a cultural system. In *Anthropological approaches to the study of religion*, M. Banton (ed.) (A.S.A. Monogr. **3**). London: Tavistock.

Godelier, M. 1972. *Rationality and irrationality in economics*. London: New Left Books.

Haldane, J.B.S. 1956. The argument from animals to men: an examination of its validity for anthropology. *J.R. anthrop. Inst.* **36**, 1–14.

Harris, M. 1968. *The rise of anthropological theory*. New York: Crowell.

——— 1971. *Culture, man and nature*. New York: Crowell.

Higgs, E.S. and M.R. Jarman 1972. The origins of animal and plant husbandry. In *Papers in economic prehistory*, E.S. Higgs (ed.). Cambridge: University Press.

Huxley, T.H. 1894. *Man's place in nature, and other anthropological essays*. London: MacMillan.

Kroeber, A.L. 1917. The superorganic. *Am. Anthrop.* **19** (2), 163–213.

Lee, R.B. 1969. !Kung Bushman subsistence: an input-output analysis. In

Environment and cultural behaviour, A.P. Vayda (ed.). Garden City: Natural History Press.

Marx, K. 1930. *Capital*. London: Dent.

—— 1964. *Pre-capitalist economic formations*, E.J. Hobsbawm (ed.). London: Lawrence and Wishart.

Meillassoux, C. 1972. From reproduction to production. *Economy and Society* **1**, 93–105.

Napier, J. 1962. The evolution of the hand. *Scientific American* **207** (6), 56–62.

Nicholson, A.J. 1960. The role of population dynamics in natural selection. In *Evolution after Darwin Vol. I: The evolution of life*, S. Tax (ed.). Chicago: University Press.

Popper, K. 1972. *Objective knowledge: an evolutionary approach*. Oxford: Clarendon.

Rappaport, R.A. 1968. *Pigs for the ancestors*. New Haven: Yale University Press.

—— 1971. Nature, culture and ecological anthropology. In *Man, culture and society* (revised edition), H.L. Shapiro (ed.). New York: Oxford University Press.

Sahlins, M.D. 1969. Economic anthropology and anthropological economics. *Soc. Sci. Inform.* **8**, (5), 13–33.

—— 1972. *Stone age economics*. London: Tavistock.

—— 1976. *The use and abuse of biology*. London: Tavistock.

Schmidt, A. 1971. *The concept of nature in Marx*. London: New Left Books.

Stern, J.T. 1970. The meaning of 'adaptation' and its relation to the phenomenon of natural selection. In *Evolutionary Biology*, Th. Dobzhansky, M.K. Hecht and E.C. Steere (eds.), **4**, 38–66. New York: Appleton-Century-Crofts.

Steward, J.H. 1955. *Theory of culture change*. Urbana: University of Illinois Press.

Terray, E. 1972. *Marxism and 'primitive' societies*. New York: Monthly Review Press.

Zohary, D. 1969. The progenitors of wheat and barley in relation to domestication and agricultural dispersal in the Old World. In *The domestication and exploitation of plants and animals*, P.J. Ucko and G.W. Dimbleby (eds.). London: Duckworth.

NOTES ON CONTRIBUTORS

Barnard, Alan. Born 1949, USA. Educated at George Washington University, B.A.; McMaster University, M.A.; University of London (University College), Ph.D.
Junior Lecturer, University of Cape Town, 1972–73; Lecturer, University College London, 1976–78; Lecturer, University of Edinburgh 1978–
    Author of 'Australian Models in the S.W.A. Highlands', *African Studies* (1975); 'Universal Systems of Kin Categorization', *African Studies* (1978); and *Bushmen* (1978).

Bonte, Pierre. Born 1942, France. Educated at the Sorbonne, Licence in sociology; University of Paris V, Doctorat d'Ethnologie.
Chargé de recherche, Centre National de la Recherche Scientifique, 1973–
    Author of *L'organisation économique des touaregs Kel Gress* (1976); 'Structures de classe et structures sociales chez les touareg Kel Gress', *Revue de l'Occident Musulman* (1976); 'Histoire et histoires: la conception du passé chez les hausa et les touareg Kel Gress' *Cahiers d'études africaines* (1977); *Segmentarité et pouvoir chez les éleveurs sahariens et sahéliens* (1978) and other books and articles.

Brown Glick, Paula. Born 1925, USA. Educated at the University of Chicago, B.A., M.A.; University of London, Ph.D.
Research Assistant and Assistant Lecturer, University College London, 1948–51; Research Anthropologist, Institute of Industrial Relations, University of California at Los Angeles, 1952–55; Lecturer, University of Wisconsin, 1955–56; Research Fellow, Fellow and Senior Fellow, The Australian National University, 1956–65; Visiting Lecturer, Faculty of Archaeology and Anthropology, Cambridge University, 1965–66; Associate Professor and Professor, State University of New York at Stony Brook, 1966–
    Co-author (with H.C. Brookfield) of *Struggle for Land* (1963) and *The People of Vila* (1969). Author of *The Chimbu* (1972), *Highland Peoples of New Guinea* (1978) and 'New Guinea: ecology, society and culture', *Annual Review of Anthropology* (1978). Co-editor (with G. Buchbinder) of *Man and Woman in the New Guinea Highlands* (1976).

Burnham, Philip. Born 1942, USA. Educated at Cornell University, B.A.;
University of California at Los Angeles, M.A., Ph.D.
Lecturer in Anthropology, University College London, 1970—; Lecturer
in Sociology, Ahmadu Bello University, 1973—74; Directeur d'études
associé, Ecole des Hautes Etudes en Sciences Sociales, 1977.
Author of 'Racial classification and identity in the Meiganga region:
North Cameroon', in P. Baxter and B. Sansom (eds.) *Race and Social
Difference* (1972); 'The explanatory value of the concept of adaptation in
studies of culture change', in C. Renfrew (ed.) *The Explanation of Culture
Change* (1973); 'Regroupment and mobile societies: two Cameroon cases',
Journal of African History (1975) and other articles.

Ellen, Roy F. Born 1947, United Kingdom. Studied at London School of
Economics, B.Sc., Ph.D.
Social Science Research Council Studentship, 1968—72; Temporary Lecturer
in Anthropology, London School of Economics, 1972—73; Lecturer in Social
Anthropology, University of Kent at Canterbury, 1973—
Author of *Nuaula Settlement and Ecology* (1978); 'The marsupial in
Nuaulu ritual behaviour', *Man* (1972); 'Variable constructs in Nuaulu zoological
classification', *Social Science Information* (1975); 'The development of
anthropology and colonial policy in the Netherlands', *Journal of the History
of the Behavioural Sciences* (1976); 'Resource and commodity', *Journal of
Anthropological Research* (1977); and other papers.

Fox, James. Born 1940. Educated at Harvard University, A.B.; Oxford University,
Diploma Social Anthrop., B. Litt., D.Phil.
Visiting Assistant Professor, Duke University, 1968—69; Assistant/Associate
Professor, Harvard University, 1969—75; Fellow, Center for Advanced Study
in the Behavioural Sciences, Stanford, 1971—72; Professorial Fellow, Research
School of Pacific Studies, Australian National University, 1975—; Fellow,
Netherlands Institute for Advanced Study, Wassenaar, 1977—78.
Author of 'Sister's child as plant', in R. Needham (ed.), *Rethinking Kin
Kinship and Marriage* (1971); 'Rotinese dynastic genealogy', in
T. Beidelman (ed.), *The Translation of Culture* (1971); 'On bad death and
the left hand', in R. Needham (ed.), *Right and Left* (1973); Our ancestors
spoke in pairs', in R. Bauman and J. Sherzer (eds.), *Explorations in the
Ethnography of Speaking* (1974); 'On binary categories and primary
symbols', in R. Willis (ed.), *The Interpretation of Symbolism* (1975); *Harvest
of the Palm* (1977).

Friedman, Jonathan. Born 1946, USA. Educated at Columbia University, B.A.,
1967; Ecole Pratique des Hautes Etudes, Institut d'Ethnologie, Sorbonne,
Licence, 1968; Columbia University, Ph.D., 1972.
Chargé de cours, Ecole Pratique des Hautes Etudes, 1972—73; Visiting
Lecturer, Institute of Social Anthropology, Uppsala, Autumn 1972;
Institute of Social Anthropology, Göteborg, 1973, 1974; Lecturer, University

College London, 1973–74; Lecturer, Institut for Etnologi og Antropologi, University of Copenhagen, 1974–

Author of 'Marxism, Structuralism, and Vulgar Materialism', *Man* (1974); *System, Structure and Contradiction in the Evolution of 'Asiatic' Social Formations* (forthcoming); 'Hypothèses sur la dynamique et les transformations tribales', *L'Homme* (1975); 'Religion as Economy and Economy as Religion' in *The Great Feast, Festschrift for K.G. Izikowitz, Ethnos* (1975); 'Tribes, States and Transformations' in M. Bloch (ed.), *Marxist Analyses and Social Anthropology* (1975); Le lieu du fétichisme et le problème des interpretations matérialistes', *La Pensée* (1975); Co-editor (with M. Rowlands) of *The Evolution of Social Systems* (1977).

Harris, David R. Born 1930, United Kingdom. Educated at Oxford, B.A., B.Litt., M.A.; University of California at Berkeley, Ph.D.
Has lectured in the Departments of Geography at University of California, Berkeley; Queen Mary College, University of London; University of New Mexico, Albuquerque; University of Toronto; Australian National University, Canberra; currently Reader in Geography at University College London.

Author of 'Alternative pathways toward agriculture', in C. Reed (ed.), *Origins of Agriculture* (1977); 'Settling down: an evolutionary model for the transformation of mobile bands into sedentary communities', in J. Friedman and M. Rowlands (eds.), *The Evolution of Social Systems* (1977); 'Subsistence strategies across Torres Strait', in J. Allen *et al.* (eds.), *Sunda and Sahul: Prehistoric Studies in Island Southeast Asia, Melanesia and Australia* (1977); 'Adaptation to a tropical rain-forest environment: Aboriginal subsistence in northeastern Queensland', in N. Blurton-Jones and V. Reynolds (eds.), *Adaptation and Behaviour* (1978); 'Group territories or territorial groups? Comments on an inter-disciplinary problem in cultural ecology', in D. Green *et al.* (eds.), *Social Organization and Settlement* (1978); *Plants, Animals, and Man in the Outer Leeward Islands, West Indies* (1965), and other books and articles.

Ingold, Tim. Born 1948, England. Studied at the University of Cambridge, B.A., Ph.D.
Visiting Lecturer, University of Helsinki, 1973–74; Lecturer in Social Anthropology, University of Manchester, 1974–

Author of 'On Reindeer and men', *Man* (1974); 'Entrepreneur and protagonist', *Journal of Peace Research* (1974); *The Skolt Lapps Today* (1976); 'The rationalization of reindeer management among Finnish Lapps', *Development and Change* (1978); 'Statistical husbandry', in J.C. Mitchell (ed.), *Numerical Techniques; Hunters, Pastoralists and Ranchers* (in press).

Peterson, Nicolas. Born 1941, England. Educated at University of Cambridge, B.A.; University of Sydney, Ph.D.
Research Officer in Northern Territory, Australian Institute of Aboriginal Studies, 1965–68; Research Fellow, Research School of Pacific Studies

Studies 1972—75; Visiting Assoc. Professor, University of New Mexico 1975. Senior Lecturer in Anthropology, Australian National University, Canberra 1975.

Author of 'Totemism yesterday', *Man* (1972); 'Hunter-gatherer territoriality: the perspective from Australia', *Am. Anthrop.* (1975) and other papers. Editor of *Tribes and Boundaries in Australia* (1977).

Riches, David. Born 1947, London. Educated at Cambridge University, B.A., M.A.; London School of Economics, Ph.D.
Lecturer in Social Anthropology, Queen's University of Belfast.

Author of 'Cash, credit and gambling in a modern Eskimo economy', *Man* (1975); and several papers on Eskimos, hunter-gatherers, social and economic change and transactional theory in *Ethnology, Queen's University Papers in Social Anthropology*, and R. Paine (ed.) *The White Arctic.*

Van Leynseele, Pierre. Born 1933, Belgium. Educated at the College for Overseas Territories, Antwerp, Licence (1958); University of California at Los Angeles, M.A. (1966).
District Adviser in the Congo (Equatorial Province), 1960—63; Research Assistant, University of California at Los Angeles, 1965—66; Research Officer at the Institute for Economic and Social Research, University of Kinshasa, 1967—73; Lecturer in Anthropology at the University of Kinshasa (1968—71); Acting Regional Director of the Institute for Scientific Research in Central Africa (IRSAC-Equateur), 1972; Research Affiliate, Afrika-Studiecentrum, Leiden, 1975—

Author of 'Une forge artisanale dans la Haute-Ngiri', *Ngonge* (1973); 'Social Anthropology and the 'Bantu expansion' (with A. Kuper), *Africa* (1978); 'Les modifications des systèmes de production chez des populations ripuaires du Haut-Zaire', *African Economic History* (1978).

AUTHOR INDEX

Note: Lists of references following individual papers are not included in this index.

SUBJECT INDEX